CAMBRIDGE LIBRARY COLLECTION

Books of enduring scholarly value

Physical Sciences

From ancient times, humans have tried to understand the workings of the world around them. The roots of modern physical science go back to the very earliest mechanical devices such as levers and rollers, the mixing of paints and dyes, and the importance of the heavenly bodies in early religious observance and navigation. The physical sciences as we know them today began to emerge as independent academic subjects during the early modern period, in the work of Newton and other 'natural philosophers', and numerous sub-disciplines developed during the centuries that followed. This part of the Cambridge Library Collection is devoted to landmark publications in this area which will be of interest to historians of science concerned with individual scientists, particular discoveries, and advances in scientific method, or with the establishment and development of scientific institutions around the world.

The Works of John Playfair

John Playfair (1748–1819) was a Scottish mathematician and geologist best known for his defence of James Hutton's geological theories. He attended the University of St Andrews, completing his theological studies in 1770. In 1785 he was appointed joint Professor of Mathematics at the University of Edinburgh, and in 1805 he was elected Professor of Natural Philosophy. A Fellow of the Royal Society, he was acquainted with continental scientific developments, and was a prolific writer of scientific articles in the *Transactions of the Royal Society of Edinburgh* and the *Edinburgh Review*. This four-volume edition of his works was published in 1822 and is prefaced by a biography of Playfair. Volume 3 includes articles on mathematics, physics, astronomy and naval tactics, revealing the range of Playfair's scientific interests.

Cambridge University Press has long been a pioneer in the reissuing of out-of-print titles from its own backlist, producing digital reprints of books that are still sought after by scholars and students but could not be reprinted economically using traditional technology. The Cambridge Library Collection extends this activity to a wider range of books which are still of importance to researchers and professionals, either for the source material they contain, or as landmarks in the history of their academic discipline.

Drawing from the world-renowned collections in the Cambridge University Library, and guided by the advice of experts in each subject area, Cambridge University Press is using state-of-the-art scanning machines in its own Printing House to capture the content of each book selected for inclusion. The files are processed to give a consistently clear, crisp image, and the books finished to the high quality standard for which the Press is recognised around the world. The latest print-on-demand technology ensures that the books will remain available indefinitely, and that orders for single or multiple copies can quickly be supplied.

The Cambridge Library Collection will bring back to life books of enduring scholarly value (including out-of-copyright works originally issued by other publishers) across a wide range of disciplines in the humanities and social sciences and in science and technology.

The Works of John Playfair

VOLUME 3

JOHN PLAYFAIR
EDITED BY JAMES G. PLAYFAIR

CAMBRIDGE UNIVERSITY PRESS

Cambridge, New York, Melbourne, Madrid, Cape Town,
Singapore, São Paolo, Delhi, Tokyo, Mexico City

Published in the United States of America by Cambridge University Press, New York

www.cambridge.org
Information on this title: www.cambridge.org/9781108029407

This edition first published 1822
This digitally printed version 2011

ISBN 978-1-108-02940-7 Paperback

THE

WORKS

OF

JOHN PLAYFAIR, ESQ.

&c. &c. &c.

Printed by George Ramsay and Company.

THE

WORKS

OF

JOHN PLAYFAIR, ESQ.

LATE

PROFESSOR OF NATURAL PHILOSOPHY IN THE UNIVERSITY OF EDINBURGH,

PRESIDENT OF THE ASTRONOMICAL INSTITUTION OF EDINBURGH,

FELLOW OF THE ROYAL SOCIETY OF LONDON,

SECRETARY OF THE ROYAL SOCIETY OF EDINBURGH,

AND HONORARY MEMBER OF THE ROYAL MEDICAL SOCIETY OF EDINBURGH.

WITH

A MEMOIR OF THE AUTHOR.

VOL. III.

EDINBURGH:

PRINTED FOR ARCHIBALD CONSTABLE & CO. EDINBURGH,

AND HURST, ROBINSON, AND CO. LONDON.

1822.

CONTENTS

OF

VOLUME THIRD.

ON

THE ARITHMETIC

OF

IMPOSSIBLE QUANTITIES.

ARITHMETIC

OF

IMPOSSIBLE QUANTITIES.*

1. THE paradoxes which have been introduced into algebra, and remain unknown in geometry, point out a very remarkable difference in the nature of those sciences. The propositions of geometry have never given rise to controversy, nor needed the support of metaphysical discussion. In algebra, on the other hand, the doctrine of negative quantities and its consequences have often perplexed the analyst, and involved him in the most intricate disputations. The cause of this diversity, in sciences which have the same object, must no doubt be sought for in the different modes which they employ to express our ideas. In geometry every magnitude is represented by one of the same kind; lines are represented by a line, and angles by an angle. The genus is always signified by the individual, and a general

* From the Transactions of the Royal Society of London, Vol. LXVIII. (1779.)—ED.

idea by one of the particulars which fall under it. By this means all contradiction is avoided, and the geometer is never permitted to reason about the relations of things which do not exist, or cannot be exhibited. In algebra again every magnitude being denoted by an artificial symbol, to which it has no resemblance, is liable, on some occasions, to be neglected, while the symbol may become the sole object of attention. It is not perhaps observed where the connection between them ceases to exist, and the analyst continues to reason about the characters after nothing is left which they can possibly express: if then, in the end, the conclusions which hold only of the characters be transferred to the quantities themselves, obscurity and paradox must of necessity ensue. The truth of these observations will be rendered evident by considering the nature of imaginary expressions, and the different uses to which they have been applied.

2. Those expressions, as is well known, owe their origin to a contradiction taking place in that combination of ideas which they were intended to denote. Thus, if it be required to divide the given line AB (fig. 1.) $=a$ in C, so that $AC \times CB$ may be equal to a given space b^2, and if $AC = x$, then $x = \frac{1}{2}a \pm \sqrt{\frac{1}{4}a^2 - b^2}$; which value of x is imaginary when b^2 is greater than $\frac{1}{4}a^2$; now to suppose that b^2 is

greater than $\frac{1}{4}a^2$, is to suppose that the rectangle AC × CB is greater than the square of half the line AB, which is impossible. The same holds wherever expressions of this kind occur. Thus, when it is asserted that unity has the three cube roots $1, \frac{-1+\sqrt{-3}}{2}, \frac{-1-\sqrt{-3}}{2}$, no more is meant than that when the general equation $x^3-ax^2+bx-r=o$ is, by a change in the data, reduced to the particular state $x^3-1=o$, x is then equal to unity only, and admits not of any other value, as it does in more general forms of the equation. The natural office of imaginary expressions is, therefore, to point out when the conditions, from which a general formula is derived, become inconsistent with each other; and they correspond in the algebraic calculus to that part of the geometrical analysis, which is usually styled the determination of problems.

3. This, however, is not the only use to which imaginary expressions have been applied. When combined according to certain rules, they have been put to denote real quantities, and though they are in fact no more than marks of impossibility, they have been made the subjects of arithmetical operations; their ratios, their products, and their sums, have been computed, and, what may seem strange, just conclusions have in that way been deduced. Nevertheless, the name of reasoning cannot be

given to a process into which no idea is introduced. Accordingly geometry, which has its modes of reasoning that correspond to every other part of the algebraic calculus, has nothing similar to the method we are now considering; for the arithmetic of mere characters can have no place in a science which is immediately conversant with ideas.

But though geometry rejects this method of investigation, it admits, on many occasions, the conclusions derived from it, and has confirmed them by the most rigorous demonstration. Here then is a paradox which remains to be explained. If the operations of this imaginary arithmetic are unintelligible, why are they not also useless? Is investigation an art so mechanical, that it may be conducted by certain manual operations? Or is truth so easily discovered, that intelligence is not necessary to give success to our researches? These are difficulties which it is of some importance to resolve, and on which much attention has not hitherto been bestowed. Two celebrated mathematicians, Bernoulli and Maclaurin, have indeed touched on this subject; but being more intent on applying their calculus, than on explaining the grounds of it, they have only suggested a solution of the difficulty, and one too by no means satisfactory. They allege,*

---

* Op. J. Bernoulli, Tom. I. No. 70. Maclaurin, Flux. Art. 699-763.

that when imaginary expressions are put to denote real quantities, the imaginary characters involved in the different terms of such expressions do then compensate or destroy each other. But, beside that, the manner in which this compensation is made, in expressions ever so little complicated, is extremely obscure, if it be considered that an imaginary character is no more than a mark of impossibility, such a compensation becomes altogether unintelligible: for how can we conceive one impossibility removing or destroying another? Is not this to bring impossibility under the predicament of quantity, and to make it a subject of arithmetical computation? And are we not thus brought back to the very difficulty to be removed? Their explanation cannot of consequence be admitted; but, on attempting another, it behoves us to observe, that a more extensive application of this method, than had been made in their time, has now greatly facilitated the inquiry. We begin then with considering the manner in which the imaginary expressions, supposed to denote real quantities, are derived; and the cases in which they prove useful for the purposes of investigation.

4. Let a be an arch of a circle, of which the radius is unity, and let c be the number which has unity for its hyperbolic logarithm, then the sine of the arch a, or

$\sin.a = \frac{c^{a\sqrt{-1}} - c^{-a\sqrt{-1}}}{2\sqrt{-1}}$; and $\cos.a = \frac{c^{a\sqrt{-1}} + c^{-a\sqrt{-1}}}{2}$.

These exponential and imaginary values of the sine and cosine are already well known to geometers; and the investigation of them, according to the received arithmetic of impossible quantities, may be as follows:

Let $\sin. a = z$, then $\dot{a} = \frac{\dot{z}}{\sqrt{1-z^2}}$. To bring this fluxion under such a form that its fluent may be found by logarithms, both numerator and denominator are to be multiplied by $\sqrt{-1}$; then

$\dot{a} = \sqrt{-1} \times \frac{\dot{z}}{\sqrt{z^2-1}}$, and (by form. 6. Harm. Men.)

$a = \sqrt{-1} \times \log. \frac{z + \sqrt{z^2-1}}{\sqrt{-1}}$. Hence $\frac{a}{\sqrt{-1}}$, or $1 \times \frac{a}{\sqrt{-1}} =$ $\log. \frac{z+\sqrt{z^2-1}}{\sqrt{-1}}$; and because 1 is the log. of c,

$c^{\frac{a}{\sqrt{-1}}} = \frac{z+\sqrt{z^2-1}}{\sqrt{-1}}$; wherefore, if both parts of the fractional index of c be multiplied by $\sqrt{-1}$,

$c^{-a\sqrt{-1}} = \frac{z+\sqrt{z^2-1}}{\sqrt{-1}}$. Again, if the arch a be considered as negative, its sine becomes also negative, and therefore $-a = \sqrt{-1} \times \log. \frac{-z+\sqrt{z^2-1}}{\sqrt{-1}}$, or, $-a\sqrt{-1}$ $= -\log. \frac{-z+\sqrt{z^2-1}}{\sqrt{-1}}$, and $a\sqrt{-1} = \log. \frac{-z+\sqrt{z^2-1}}{\sqrt{-1}}$;

whence also, $c^{a\sqrt{-1}} = \frac{-z+\sqrt{z^2-1}}{\sqrt{-1}}$. If from this equation the former be taken away, there remains $-\frac{2z}{\sqrt{-1}}=c^{a\sqrt{-1}}-c^{-a\sqrt{-1}}$, whence dividing by $2\sqrt{-1}$ we have $z=\sin.a=\frac{c^{a\sqrt{-1}}-c^{-a\sqrt{-1}}}{2\sqrt{-1}}$. By adding together the equations, a value of the cosine may be found in the same imaginary terms which were assigned above. Now, by means of these expressions many theorems may be demonstrated; it may, for example be shown, that if a and b are any two arches of a circle, of which the radius is unity, then $\sin.a\times\cos.b=\sin.\frac{1}{2}(a+b)+\sin.\frac{1}{2}(a-b)$. For

$\sin.a=\frac{c^{a\sqrt{-1}}-c^{-a\sqrt{-1}}}{2\sqrt{-1}}$, and $\cos.b=\frac{c^{b\sqrt{-1}}+c^{-b\sqrt{-1}}}{2}$, therefore, $\sin.a\times\cos.b=$

$$\frac{c^{(a+b)\sqrt{-1}}-c^{(-a-b)\sqrt{-1}}+c^{(a-b)\sqrt{-1}}-c^{(b-a)\sqrt{-1}}}{4\sqrt{-1}}=$$

$\sin.\frac{1}{2}(a+b)+\sin.\frac{1}{2}(a-b)$.

5. Now it may be observed, that the imaginary value which has been found for $\sin.a$ was obtained by bringing a fluxion properly belonging to the circle, under the form of one belonging to the hyperbola. It may, therefore, be worth while to inquire, whether a similar expression may not be derived from the hyperbola itself.

Let BAD be a rectangular hyperbola, (fig. 2,) of which the centre is C, and the semi-transverse axis

AC=1; let B be any point in the hyperbola, join BC, and let BE be an ordinate to the transverse axis. Then, if the sector ACB$=\frac{1}{2}a$, and BE$=z$, it is known that $a=\frac{z}{\sqrt{1+z^2}}$; whence $a=\log(z+\sqrt{1+z^2})$, and $c^a=z+\sqrt{1+z^2}$. But if the sector be taken on the other side of the transverse axis, a and z become negative, and $c^{-a}=-z+\sqrt{1+z^2}$. Hence $z=\frac{c^a-c^{-a}}{2}$; in like manner the abscissa belonging to ACB, that is CE$=\frac{c^a+c^{-a}}{2}$. These values of the ordinates and abscissæ differ in nothing from those of the sines and cosines already found, except in being free from impossible quantities; for it is evident, that the quantity a is related in the same manner both to the circular and hyperbolic sectors. If now ord.a and abs b denote the ordinate and abscissa belonging to the sectors $\frac{1}{2}a$, $\frac{1}{2}b$ respectively,

$$\text{ord } a \times \text{abs.} b = \left(\frac{c^a-c^{-a}}{2}\right)\left(\frac{c^b+c^{-b}}{2}\right)=$$

$$\frac{c^{a+b}-c^{-a-b}+c^{a-b}-c^{b-a}}{4}=\frac{\text{ord.}(a+b)}{2}+\frac{\text{ord.}(a-b)}{2}.$$

6. The conclusions in both the foregoing cases are perfectly coincident, and the methods by which they have been obtained are similar; though with this difference between them, that in the first all the steps are unintelligible, but in the last signifi-

cant. If then, notwithstanding a difference which might be expected so materially to affect their conclusions, they have been equally successful in the discovery of truth, it can be ascribed only to the analogy which takes place between the subjects of investigation; an analogy so close, that every property belonging to the one may, with certain restrictions, be transferred to the other. Accordingly, every imaginary expression which has been found to belong to the circle in the preceding calculation, is by the substitution of real for impossible quantities, or of $\sqrt{1}$ for $\sqrt{-1}$, converted into a proposition which holds of the hyperbola. The operations, therefore, performed with the imaginary characters, though destitute of meaning themselves, are yet notes of reference to others which are significant. They point out, indirectly, a method of demonstrating a certain property of the hyperbola, and then leave us to conclude from analogy, that the same property belongs also to the circle. All that we are assured of by the imaginary investigation is, that its conclusion may, with all the strictness of mathematical reasoning, be proved of the hyperbola; but if from thence we would transfer that conclusion to the circle, it must be in consequence of the principle which has been just now mentioned. The investigation, therefore, resolves itself ultimately into an argument from analogy; and, after the strictest examination, will be found without any other claim to the evidence of demonstration. Had the foregoing proposition been proved of the hyperbola only, and afterwards concluded to hold of the circle, merely from the affinity of the curves, its certainty would have been precisely the same as when a proof is made out by the intervention of imaginary symbols.

7. Let AB, AC, AD, AE, (fig. 3,) be any arches of a circle in arithmetical progression, and let m be their number; it is required to find the sum of the sines BG, CH, &c. of those arches. Let the radius AF=1, AB=a, and the common difference of the arches, or BC=x; the sum of the series $\sin.a+\sin(a+x)+\sin(a+2x)+\ldots\ldots m)$ is to be found. Now, because

$\sin.a=\frac{c^{a\sqrt{-1}}-c^{-a\sqrt{-1}}}{2\sqrt{-1}}$, and $\sin.(a+x)=$
$\frac{c^{(a+x)\sqrt{-1}}-c^{(-a-x)\sqrt{-1}}}{2\sqrt{-1}}$; the series $\sin.a+\sin(a+x)$
$+\sin(a+2x)\ldots\ldots m)=\frac{c^{a\sqrt{-1}}}{2\sqrt{-1}}\left(1+c^{x\sqrt{-1}}+c^{2x\sqrt{-1}}\ldots\ldots\right.$
$\left.\ldots m\right)-\frac{c^{a\sqrt{-1}}}{2\sqrt{-1}}\left(1+c^{-x\sqrt{-1}}+c^{-2x\sqrt{-1}}\ldots\ldots\ldots m\right)$.

But these series are both geometrical progressions, and the sum of the first is $\left(\frac{c^{a\sqrt{-1}}}{2\sqrt{-1}}\right)\left(\frac{1-c^{mx\sqrt{-1}}}{1-c^{x\sqrt{-1}}}\right)$;

and of the second, $\left(\frac{c^{-a\sqrt{-1}}}{2\sqrt{-1}}\right)\left(\frac{1-c^{-mx\sqrt{-1}}}{1-c^{-x\sqrt{-1}}}\right)$. The sum of the proposed series therefore =

$$\left(\frac{c^{a\sqrt{-1}}}{2\sqrt{-1}}\right)\left(\frac{1-c^{mx\sqrt{-1}}}{1-c^{x\sqrt{-1}}}\right)-\left(\frac{c^{-a\sqrt{-1}}}{2\sqrt{-1}}\right)\left(\frac{1-c^{-mx\sqrt{-1}}}{1-c^{-x\sqrt{-1}}}\right)$$

$$=\left(\frac{1}{2\sqrt{-1}}\right)\left(\frac{c^{a\sqrt{-1}}-c^{(a+mx)\sqrt{-1}}-c^{(a-x)\sqrt{-1}}+c^{(a+mx-x)\sqrt{-1}}}{1-c^{x\sqrt{-1}}-c^{-x\sqrt{-1}}+1}\right)+$$

$$\left(\frac{1}{2\sqrt{-1}}\right)\left(\frac{-c^{-a\sqrt{-1}}+c^{(-a-mx)\sqrt{-1}}+c^{(-a+x)\sqrt{-1}}-c^{(-a-mx+x)\sqrt{-1}}}{1-c^{x\sqrt{-1}}-c^{-x\sqrt{-1}}+1}\right);$$

in which expression, if the sines be substituted for their imaginary values, we have

$$\frac{\sin.a-\sin(a+mx)-\sin(a-x)+\sin(a+mx-x)}{2(1-\cos.x)}=\sin.a+$$
$$\sin(a+x)+\sin(a+2x)\ldots\ldots\ldots m). \quad \text{Q.E.I.}$$

When AB=BC, or $a=x$, the proposed series becomes $\sin.x+\sin.2x+\sin.3x\ldots.m$, and its value $=\frac{\sin.x-\sin(m+1)x+\sin.mx}{2(1-\cos.x)}$.

In like manner it will be found, that the sum of the cosines of the same arches, or

$$\cos.a+\cos(a+x)+\cos(a+2x)\ldots\ldots.m=$$
$$\frac{\cos.a-\cos(a+mx)-\cos(a-x)+\cos(a+mx-x)}{2(1-\cos.x)};$$ and when

$$a=x,\ \cos.x+\cos.2x+\cos.3x\ldots\ldots m=\frac{\cos.mx-\cos(m+1)x}{2(1-\cos.x)}$$
$$-\frac{1}{2}.$$

8. To solve the same problem, in the case of the hyperbola, we must follow the steps which have been traced out by these imaginary operations. Let ABE be an equilateral hyperbola, (fig.

4,) of which the centre is F, and the semitransverse axis AF=1; let ABF, ACF, ADF, &c. be any sectors in arithmetical progression, and let m be their number, it is required to find the sum of all the ordinates BG, CH, DK, &c. belonging to those sectors. Let the sector AFB$=\frac{1}{2}a$, and the sector BFC, which is the common difference of the sectors, $=\frac{1}{2}x$: then BG, or ord.$a=\frac{c^a-c^{-a}}{2}$, and CH, or ord.$(a+x)=\frac{c^{(a+x)}-c^{(-a-x)}}{2}$, by Art. 5. Therefore the series of ordinates, that is, BG+CH +DK+ ...... $m=\frac{c^a}{2}\left(1+c^x+c^{2x}+\ldots\ldots m\right)-$ $\frac{c^{-a}}{2}\left(1+c^{-x}+c^{-2x}+\ldots\ldots m\right)=\frac{c^a}{2}\left(\frac{1-c^{mx}}{1-c^x}\right)-$ $\frac{c^{-a}}{2}\left(\frac{1-c^{-mx}}{1-c^{-x}}\right)=$

$$\frac{1}{2}\left(\frac{c^a-c^{(a+mx)}-c^{(a-x)}+c^{(a+mx-x)}-c^{-a}+c^{(-a-mx)}+c^{(-a-x)}-c^{(-a-mx+x)}}{1-c^x-c^{-x}+1}\right)$$

$$=\frac{\text{ord.}a-\text{ord.}(a+mx)-\text{ord.}(a-x)+\text{ord.}(a+mx-x)}{2(1-\text{abs.}x)}$$

When $a=x$, ord.x+ord.$2x$+ord.$3x+\ldots\ldots m=$ $\frac{\text{ord.}x-\text{ord.}(m+1)x+\text{ord.}mx}{2(1-\text{abs.}x)}$

In like manner it is proved, that the sum of the abscissæ, that is FG+FH+FK+..... $m=$ $\frac{\text{abs.}a-\text{abs.}(a+mx)-\text{abs.}(a-x)+\text{abs.}(a+mx-x)}{2(1-\text{abs.}x)}$; and

when $a=x$, this expression becomes
$\frac{\text{abs.}mx-\text{abs.}(m+1)x}{2(1-\text{ab}.x)}-\frac{1}{2}$.

9. The coincidence of the theorems deduced in the two last articles is obvious at first sight, and if the methods by which they have been obtained be compared, it will appear, that the imaginary operations in the one case were of no use but as they adumbrated the real demonstration, which took place in the other. This will be rendered more evident by considering that the resolution of the series of hyperbolic ordinates, into two others of continual proportionals, can be exhibited geometrically. For, from the points A, B, C, and D, let AM, BN, CO, DP, be drawn at right angles to the asymptote FP; let GB produced meet FP in Q, and let BR be perpendicular to the conjugate axis FR. Then, because the triangles FRS, FMA, are equiangular, AF : FM :: FS : FR; hence $FR=\frac{FM}{FA}\times FS=\frac{FM}{FA}(FN-NB)$. For the same reason $CH=\frac{FM}{FA}(FO-OC)$, and $DR=\frac{FM}{FA}(FP-PD)$. Therefore $BG+CH+DK=\frac{FM}{FA}(FN+FO+FP)-\frac{FM}{FA}(BN+CO+DP)$; now, FN, FO, FP, are continual proportionals, and so also are BN, FO, FP, because the sectors FBC, FCD, are equal. But in the circle no such resolution of the proposed series of

sines can take place, that series being subject to alternate increase and diminution; on which account it is, that imaginary characters enter into the exponential value of the sine. Those characters are, therefore, so far from compensating each other in the present case, as they ought to do, on the supposition of Bernoulli and Maclaurin, that they manifestly serve as marks of impossibility. There remains, of consequence, the affinity between circular arches and hyperbolic areas, or between the measures of angles and of ratios, as the only principle on which the imaginary investigation can proceed. It need scarcely be observed, that the exponential value of the hyperbolic ordinate may be deduced from what has been proved in this article.

10. But as the arithmetic of impossible quantities is no where of greater use than in the investigation of fluents, it is of consequence to inquire, whether the preceding theory extends also to that application of it.

Let it then be required to find the fluent of the equation $\frac{\ddot{y}}{\dot{x}^2}=a^2y=Q$, where Q denotes any function whatever of x. For this purpose, the following lemma is premised: let x be any arch, and p any flowing quantity; then, if the sign $\int$, be taken to denote the fluent of the quantity to which it is prefixed, $\sin.x\int\dot{p}\cos.x - \cos.x\int\dot{p}\sin.x = \frac{c^{x\sqrt{-1}}}{2\sqrt{-1}}\int\dot{p}c^{-x\sqrt{-1}}$ $-\frac{c^{-x\sqrt{-1}}}{2\sqrt{-1}}\int\dot{p}c^{x\sqrt{-1}}$; or if $\frac{1}{2}x$ be a hyperbolic sector, $\text{ord}.x\int\dot{p}\,\text{abs}.x - \text{abs}.x\int\dot{p}\,\text{ord}.x = \frac{c^x}{2}\int\dot{p}c^{-x} - \frac{c^{-x}}{2}\int\dot{p}c^x$.

Because $\sin.x\int\dot{p}\cos.x = \frac{c^{x\sqrt{-1}} - c^{-x\sqrt{-1}}}{2\sqrt{-1}}\int\dot{p}\times$ $\frac{c^{x\sqrt{-1}} + c^{-x\sqrt{-1}}}{2}$, by separating the terms we have $\sin.x\int\dot{p}\cos.x = \frac{c^{x\sqrt{-1}}}{4\sqrt{-1}}\int\dot{p}c^{x\sqrt{-1}} + \frac{c^{x\sqrt{-1}}}{4\sqrt{-1}}\int\dot{p}c^{-x\sqrt{-1}} -$ $\frac{c^{-x\sqrt{-1}}}{4\sqrt{-1}}\int\dot{p}c^{x\sqrt{-1}} - \frac{c^{-x\sqrt{-1}}}{4\sqrt{-1}}\int\dot{p}c^{-x\sqrt{-1}}$, for the same reason, $-\cos.x\int\dot{p}\sin.x = -\frac{c^{x\sqrt{-1}}}{4\sqrt{-1}}\int\dot{p}c^{x\sqrt{-1}} +$ $\frac{c^{x\sqrt{-1}}}{4\sqrt{-1}}\int\dot{p}c^{-x\sqrt{-1}} - \frac{c^{-x\sqrt{-1}}}{4\sqrt{-1}}\int\dot{p}c^{x\sqrt{-1}} + \frac{c^{-x\sqrt{-1}}}{4\sqrt{-1}}\times$ $\int\dot{p}c^{-x\sqrt{-1}}$. Wherefore, by collecting the sum of all the terms, we have $\sin.x\int\dot{p}\cos.x - \cos.x\int\dot{p}\sin.x =$ $\frac{c^{x\sqrt{-1}}}{2\sqrt{-1}}\int\dot{p}c^{-x\sqrt{-1}} - \frac{c^{-x\sqrt{-1}}}{2\sqrt{-1}}\int\dot{p}c^{x\sqrt{-1}}$. The demonstration in the case of the hyperbola is free from imaginary expressions; but, in other respects, is exactly similar to that which has now been given in the case of the circle.

11. Let the coefficient of y, in the proposed equation, be first supposed negative, that is, let $\frac{\ddot{y}}{\dot{x}^2} - a^2 y = Q$, and if we multiply by $c^{nx}\dot{x}$, n being a constant but indeterminate quantity, it becomes $\frac{c^{nx}\ddot{y}}{\dot{x}} - a^2 c^{nx} y\dot{x} = c^{nx} Q\dot{x}$. Let $c^{nx}\left(\frac{A\dot{y}}{\dot{x}} - By\right)$ be assumed for the fluent, A and B being indeterminate, and let its fluxion be taken, then,

$$\frac{Ac^{nx}\ddot{y}}{\dot{x}} + nAc^{nx}\dot{y} - nBc^{nx}y\dot{x} = c^{nx}Q\dot{x}.$$
$$-Bc^{nx}\dot{y}.$$

Hence, by comparing the terms, we get $A=1$, $nA-B=0$, $nB=a^2$; therefore, $n=\pm a$, and $B=\pm a$: for n and B let the value $+a$ be substituted, and for A, its value, unity; and the assumed equation becomes $\left(\frac{\dot{y}}{\dot{x}} - ay\right)c^{ax} = \int c^{ax}Q\dot{x}$, or, $\frac{\dot{y}}{\dot{x}} - ay = c^{-ax}\int c^{ax}Q\dot{x}$. Let this equation be multiplied by $c^{mx}\dot{x}$, m being indeterminate as before, and $c^{mx}\dot{y} - ac^{mx}y\dot{x} = c^{(m-a)x}\dot{x}\int c^{ax}Q\dot{x}$. The fluent of the first member of this equation is evidently of the form $Dc^{mx}y$, the fluxion of which, viz. $Dc^{mx}\dot{y} + Dmc^{mx}y\dot{x}$ being compared with the former, gives $D=1$, and $m=-a$; wherefore, $c^{-ax}y = \int c^{-2ax}\dot{x}\int c^{ax}Q\dot{x}$, or $y = c^{ax}\int c^{-2ax}\dot{x}\int c^{ax}Q\dot{x}$. Let $c^{ax}Q\dot{x} = \dot{z}$, and $c^{-2ax}\dot{x} = \dot{v}$;

then $\int c^{-2ax}\dot{x}\int c^{ax}Q\dot{x} = \int z\dot{v} = zv - \int v\dot{z}$; but $v = \frac{1}{2a} - \frac{c^{-2ax}}{2a}$, supposing that v and x vanish at the same time; therefore, $vz - \int v\dot{z} = \frac{1}{2a}\int c^{ax}Q\dot{x} - \frac{c^{-2ax}}{2a}\int c^{ax}Q\dot{x} - \frac{1}{2a}\int c^{ax}Q\dot{x} + \frac{1}{2a}\int c^{-ax}Q\dot{x} = \frac{1}{2a}\int c^{-ax}Q\dot{x} - \frac{c^{-2ax}}{2a}\int c^{ax}Q\dot{x}$. Hence $y = \frac{c^{ax}}{2a}\int c^{-ax}Q\dot{x} - \frac{c^{-ax}}{2a}\int c^{ax}Q\dot{x}$. This value of y is sufficient for the construction of the fluent, because the quantities $\int c^{-ax}Q\dot{x}$, and $\int c^{ax}Q\dot{x}$ depend on the quadrature of the hyperbola; but if we would introduce into it the ordinates and abscissæ of that curve, we need only have recourse to the foregoing lemma, from which it appears, that $y = \frac{1}{a}\text{ord.}ax\int Q\dot{x}\text{ abs. }ax - \frac{1}{a}\text{abs.}ax\int Q\dot{x}\text{ ord.}ax$.

12. Let the coefficient of y be now supposed affirmative, or let $\frac{\ddot{y}}{\dot{x}^2} + a^2y = Q$. In this case imaginary expressions are introduced into the fluent, and the construction by the hyperbola becomes impossible. For we have then, $n = \pm a\sqrt{-1}$, from which, by proceeding as above, we get $y = \frac{c^{-ax\sqrt{-1}}}{2a\sqrt{-1}}\int c^{-ax\sqrt{-1}}Q\dot{x} -$

$\frac{c^{-ax\sqrt{-1}}}{2a\sqrt{-1}}\int c^{ax\sqrt{-1}}Q\dot{x}$; hence also, by the lemma, $y =$ $\sin. ax\int Q\dot{x}\cos. ax - \cos. ax\int Q\dot{x}\sin. ax$. Here the quantities $\int Q\dot{x}\cos. ax$, and $\int Q\dot{x}\sin. ax$, are assignable by the quadrature of the circle, in the same manner as $\int Q\dot{x}\,\text{abs.}\,ax$, and $\int Q\dot{x}\,\text{ord.}\,ax$, by the quadrature of the hyperbola; but the method of investigating them, though an illustration of the principles which we have laid down, is too well known to need to be inserted here. In like manner might the fluents of innumerable fluxionary equations, comprehended under the general form $Q = y + \frac{a\dot{y}}{\dot{x}} + \frac{b\ddot{y}}{\dot{x}^2} + \frac{d\dddot{y}}{\dot{x}^3} +$ &c. be deduced, and all of them would tend to prove that the arithmetic of impossible quantities is no more than a method of tracing the analogy between the measures of ratios and angles. Euler * and D'Alembert † were the first to integrate such equations as the preceding, and the method employed here differs from theirs only by being better adapted to illustrate the principle which is common to them all.

13. The forms in the Harmonia Mensurarum might also be brought to confirm this theory; but

---

* Nov. Comm. Petrop. Tom. III.

† Théorie de la Lune.

without accumulating instances any farther, it may be sufficient to remark two consequences that follow from it: 1. That the only cases in which imaginary expressions may be put to denote real quantities, are those in which the measures of ratios or of angles are concerned. 2. That the property of either of those measures, so investigated, might have been inferred from analogy alone. Now both these conclusions are agreeable to experience. It does not appear, that any instance has yet occurred where imaginary characters serve to express real quantities, if circular arches or hyperbolic areas are not the subjects of investigation; and if the conclusion obtained may not be transferred from the one to the other, by a mere substitution of corresponding magnitudes; that is, of sines for ordinates, cosines for abscissæ, and circular arches for the doubles of hyperbolic sectors. The affinity between the circle and hyperbola is not however so close, but that it is subject to certain limitations, from considering which, the truth of what is here asserted will be rendered more evident.

1. Any proposition demonstrated of hyperbolic sectors may be transferred to circular arches by substitution alone, without any change in the signs, when only abscissæ and their products enter into the enunciation, and conversely. Thus $\text{abs.}a \times \text{abs.}b = \frac{\text{abs}(a+b)}{2} + \frac{\text{abs}(a-b)}{2}$; and $\cos a \times \cos b = \frac{\cos(a+b)}{2}$

$+\frac{\cos(a-b)}{2}$. The same holds when the simple power of the ordinate is combined with any power whatever of the abscissa; so in the theorems of articles 5 and 4 $\text{ord}\,a\times\text{abs}\,b=\frac{\text{ord}\,(a+b)}{2}+\frac{\text{ord}\,(a-b)}{2}$; and $\sin a\times\cos b=\frac{\sin(a+b)}{2}+\frac{\sin(a-b)}{2}$.

2. When an expression containing any property of hyperbolic sectors, involves in it the rectangle of two ordinates, the value of that rectangle must have a contrary sign, when a transition is made to the circle. Thus $\text{ord}\,a\times\text{ord}\,b=\frac{\text{abs}(a+b)}{2}-\frac{\text{abs}(a-b)}{2}$; but $\sin a\times\sin b=-\frac{\cos(a+b)}{2}+\frac{\cos(a-b)}{2}$. The difference which, according to this rule, is found between the powers of ordinates and of sines may be seen in the following examples. If $\frac{1}{2}x$ denote any hyperbolic sector, then by involving $\frac{c^x-c^{-x}}{2}$, and again substituting for the exponential quantities, as in Art. 5, we have,

$$(\text{ord}\,x)^2=\frac{\text{abs}\,2x-1}{2};$$

$$(\text{ord}\,x)^3=\frac{\text{ord}\,3x-3\,\text{ord}\,x}{4};$$

$$(\text{ord}\,x)^4=\frac{\text{abs}\,4x-4\,\text{abs}\,2x+3}{8};$$

$(\text{ord}x)^5 = \frac{\text{ord}5x - 5\text{ord}3x + 10\text{ord}x}{16}$; and universally, if n be any number; A the coefficient of the second term of a binomial raised to the power n, B the coefficient of the third, &c. and p the greatest coefficient: when n is an even number,

$$(\text{ord}\,x)^n = \frac{\text{abs}nx - \text{A}\,\text{abs}(n-2)x + \text{B}\,\text{abs}(n-4)x \ldots\ldots \mp p}{2^{n-1}} \pm$$

$\frac{p}{2^n}$; but when n is an odd number,

$$(\text{ord}x)^n = \frac{\text{ord}nx - \text{A}\,\text{ord}(n-2)x + \text{B}\,\text{ord}(n-4)x \ldots\ldots \mp p\,\text{ord}x}{2^{n-1}}.$$

If now x denote an arch of a circle, by substituting and changing the signs as oft as $(\text{ord}\,x)^2$ occurs in any of the preceding expressions, we get

$$(\sin x)^2 = \frac{1 - \cos 2x}{2};$$

$$(\sin x)^3 = \frac{3\sin x - \sin 3x}{4};$$

$$(\sin x)^4 = \frac{3 - 4\cos 2x + \cos 4x}{8};$$

$(\sin x)^5 = \frac{10\sin x - 5\sin 3x + \sin 5x}{16}$; and universally, if n be any number, p the greatest coefficient of a binomial raised to the power n, A the coefficient next less than p, B the coefficient next less than A, and so on: when n is an even number,

$$(\sin x)^n = \frac{\frac{1}{2}p - \text{A}\cos 2x + \text{B}\cos 4x - \&\text{c.}}{2^{n-1}};$$

but when n is an odd number,

$$(\sin x)^n = \frac{p\sin x - \text{A}\cos 3x + \text{B}\cos 5x - \&\text{c.}}{2^{n-1}}$$

These series differ from the former only in the signs, and the arrangement of the terms; and when either n, or $n-1$, is divisible by 4, the signs remain the same in both.

14. The reason of the foregoing rule for changing the signs is, that the rectangle under two ordinates to the hyperbola is always expressed by the difference of two abscissæ; and that if from the abscissa belonging to a greater sector, be subtracted the abscissa belonging to a less, the remainder will be affirmative; whereas, if from the cosine of a greater arch be subtracted the cosine of a less, the remainder will be negative. Therefore, that the rectangles, expressed by these remainders, may have the same sign, in both cases, the signs of the remainders must be different.

It appears then, that the second rule, as well as the first, is founded on the principle of analogy when taken with the necessary limitations, and it is likewise evident from the instances which have been produced, that those rules lead to the very same conclusions which are obtained from the imaginary values of the sine and cosine.

There are, however, instances in which the analogy between the circular and hyperbolic areas being wholly interrupted, neither the foregoing rules, nor any of the same kind, can be applied; but this occasions no ambiguity, for the construction required in such cases is by its nature restricted to one of the curves only. Of this kind is the Cotesian theorem, which requires the whole circle to be divided into a given number of equal parts, and therefore cannot be extended to the hyperbola where a similar division is impossible. Others of a like nature may be derived from the general theorems already investigated; for the circle, by returning into itself, often reduces them to a simplicity to which there is nothing analogous in the hyperbola. Many examples of this might be adduced, but the two following may suffice.

1. Let ABCDE (fig. 5.) be a regular polygon inscribed in a circle, and let m be the number of its sides; it is required to find the sum of the lines FA, FB, FC, &c. drawn from any point F in the circumference, to all the angles of the polygon. By the method which in article 7, was employed to obtain the sum of the sines of a series of arches in arithmetical progression, it will be found, that the sum of the chords of the arches a, $(a+x)$, $(a+2x)$, ... (m), that is, (making $\text{FA}=a$, and $\text{AB}=x$) the sum of the chords of the arches FA, FB, FC, &c. =

$$\frac{\text{cho.}a-\text{cho}\,(a+mx)-\text{cho.}(a-x)+\text{cho.}(a+mx-x)}{2(1-\cos.\frac{1}{2}x)}\text{; but,}$$

in the present case, mx is equal to the circumference, and therefore $-\text{cho.}(a+mx)=+\text{cho.}a$ (the chord of an arch greater than the circumference being negative); and, for the same reason, cho.$(a+$

$mx-x) = -\text{cho.}(a-x) = +\text{cho.}(x-a)$. Hence the general expression becomes

$\frac{\text{cho.}a+\text{cho.}(x-a)}{1-\cos.\frac{1}{2}x} = \text{FA}+\text{FB}+\text{FC}+\ldots\ldots m.$ If therefore GK be drawn from the centre, bisecting the chord AB in H, and meeting the circumference in K, the sum of the chords, that is,

$$\text{FA}+\text{FB}+\text{FC}+\text{FD}+\text{FE} = \frac{\text{AF}+\text{FE}}{\text{HK}} \times \text{GK}.$$

2. Let n be an even number, the rest remaining as above, and let it be required to find the sum of the n powers of the chords, that is, the sum of $(\text{FA})^n+(\text{FB})^n+(\text{FC})^n \ldots\ldots m$. By reasoning, as in the case of the sines, it will appear, that if p be the greatest coefficient of a binomial raised to the power n; A the coefficient next less than p; B the coefficient next less than A, and so on, then,

$$\left(\text{cho.}a\right)^n = p - 2\text{A}\cos.a + 2\text{B}\cos.2a + 2\text{D}\cos.3a + \text{\&c.}$$

$$\left(\text{cho.}(a+x)\right)^n = p - 2\text{A}\cos.(a+x) + 2\text{B}\cos.2(a+x) + 2\text{D}\cos.3(a+x) + \text{\&c.}$$

$$\left(\text{cho.}(a+2x)\right)^n = p - 2\text{A}\cos(a+2x) + 2\text{B}\cos.2(a+2x) + 2\text{D}\cos.3(a+2x) + \text{\&c.}$$

Each of these vertical columns is to be continued downward, till the number of terms be equal to m, and therefore the sum of the second is mp. The sum of the third, or of

$$-2\text{A}\left(\cos.a+\cos.(a+x)+\cos.(a+2x)\ldots\ldots m\right),$$

by article 7, is

$$-2A\times\frac{\cos.a-\cos.(a+mx)-\cos(a-x)+\cos(a+mx-x)}{2(1-\cos.x)}=$$

(because $mx=$ the circumference),

$-A\times\frac{\cos.a-\cos.a-\cos.(a-x)+\cos.(a-x)}{1-\cos.x}=0.$ In like manner do the sums of all the subsequent columns vanish; and therefore, $\left(\text{cho.}\,a\right)^n+\left(\text{cho.}(a+x)\right)^n$ $+\left(\text{cho.}(a+2x)\right)^n\ldots\ldots m=mp.$ But when n is an even number, $p=\frac{\frac{1}{2}n+1}{\frac{1}{2}n-1}\times\frac{\frac{1}{2}n+2}{\frac{1}{2}n-2}\ldots\ldots\times\frac{n}{\frac{1}{2}n}=\frac{1.3.5.7\ldots.(n-1)}{1.2.3.4\ldots.\frac{1}{2}n}$ $\times 2^{\frac{1}{2}n}$. If therefore the radius be put $=r$, and the expression made homogeneous, we have

$(FA)^n+(FB)^n+(FC)^n\ldots\ldots.m=m\times\frac{1.3.5.7\ldots\ldots(n-1)}{1.2.3.4\ldots\ldots\frac{1}{2}n}\times$ $2^{\frac{1}{2}n}r^n$. Q. E. I.

This last coincides with the forty-first of the curious and difficult propositions published by Dr Stewart, under the title of General Theorems. It is given there without a demonstration, but appears plainly to have been investigated, in a manner altogether rigorous, by that profound geometer. It may therefore be regarded as one of the instances, in which the conclusions of the imaginary arithmetic are verified by the geometrical analysis.

15. The two foregoing propositions being confined to the circle, and yet having been investigated by the help of imaginary expressions, may, at

first sight, seem exceptions to the rule which we have been endeavouring to establish. But it needs only to be remarked, that they are particular cases of certain theorems belonging both to the circle and hyperbola, and that it was into the investigation of those theorems, that the imaginary expressions were introduced.

The conclusions therefore from the whole are these: that imaginary expressions are never of use in investigation but when the subject is a property common to the measures both of ratios and of angles; that they never lead to any consequence which might not be drawn from the affinity between those measures; and that they are indeed no more than a particular method of tracing that affinity. The deductions into which they enter are thus reduced to an argument from analogy, but the force of them is not diminished on that account. The laws to which this analogy is subject; the cases in which it is perfect, in which it suffers certain alterations, and in which it is wholly interrupted, are capable, as may be concluded from the specimens above, of being precisely ascertained. Supported on so sure a foundation, the arithmetic of impossible quantities will always remain an useful instrument in the discovery of truth, and may be of service when a more rigid analysis can hardly be applied. For this reason, many researches concerning it, which in themselves might be deemed absurd, are nevertheless not destitute of utility. Bernoulli has found, for example, that if r be the radius of a circle, the circumference $= \frac{4 \log. \sqrt{-1}}{\sqrt{-1}} r$; and the same may be deduced from article 4. Considered as a quadrature of the circle, this imaginary theorem is wholly insignificant, and would deservedly pass for an abuse of calculation; at the same time we learn from it, that if in any equation the quantity $\frac{\log. \sqrt{-1}}{\sqrt{-1}}$ should occur, it may be made to disappear, by the substitution of a circular arch, and a property, common to both the circle and hyperbola, may be obtained. The same is to be observed of the rules which have been invented for the transformation and reduction of impossible quantities:* they facilitate the operations of this imaginary arithmetic, and thereby lead to the knowledge of the most beautiful and extensive analogy which the doctrine of quantity has yet exhibited.

* The rules chiefly referred to are those for reducing the impossible roots of an equation to the form $A + B\sqrt{-1}$.

FINIS.

CAUSES

WHICH AFFECT THE ACCURACY

OF

BAROMETRICAL MEASUREMENTS.

ON

THE ACCURACY

OF

BAROMETRICAL MEASUREMENTS.*

THOUGH the labours of M. Deluc, and of the excellent observers who followed him, have brought the barometrical measurement of heights to very great exactness, they have not yet given to it the utmost perfection it can attain. Some causes of inaccuracy are still involved in it; of which we ought, at least, to estimate the effects, if we cannot correct them altogether. The allowance made on account of the temperature of the air, implies in it a hypothesis that has not been examined, nor even expressed; and many other circumstances that affect the density of the atmosphere, have either been wholly omitted, or improperly introduced.

* From the Transactions of the Royal Society of Edinburgh, Vol. I. (1788.)—ED.

The object of the present paper is to correct the errors that arise from these causes, or, where that cannot be done, to assign the limits within which those errors are contained.

1. The most important correction introduced by M. Deluc, is that which depends on the temperature of the air. His observations led him to conclude, that, at a certain temperature, marked nearly by $69\frac{1}{4}°$ of Fahrenheit, the difference of the logarithms of the heights of the mercury in the barometer, at the upper and the lower stations, gave the height of the former of those stations above the latter in 1000ths of a French toise; but that at every other temperature above or below $69\frac{1}{4}°$; a correction of .00223 of the whole was to be added or subtracted for every degree of the thermometer. By observations still more accurate, it has been found, that the temperature at which the difference of the logarithms gives the height in English fathoms, is 32°; and that the correction at other temperatures is 00243 of that difference, for every degree of the thermometer.* The man-

---

* General Roy makes the fixed temperature 32°, and the expansion for 1°, =.00245, at a medium Sir G. Shuckburgh makes the fixed temperature $31\frac{1}{4}°$, and the expansion, as here assigned, viz. .00243. *Phil. Trans.* 1777. It is sufficient for us at present to know these numbers nearly. According to the formula laid down hereafter, they will all require to be corrected.

ner of estimating the temperature of the air, adopted in all these observations, was the same ; an arithmetical mean was taken between the heights of the thermometers, at the upper and lower stations, and was supposed to be uniformly diffused through the column of air intercepted between them. M. Deluc, however, was sensible that this supposition was inaccurate; and General Roy, too, has observed, that " one of the chief causes of error in barometrical computations proceeds from the mode of estimating the temperature of the column of air from that of its extremities, which must be faulty in proportion as the height and difference of temperature are great."* It will appear, however, that this estimation, though adopted merely on account of its simplicity, and probably on no other principle than the general one of taking a mean between two observations, which, taken singly, are inaccurate, comes nearer to the truth than there was any reason to expect.

2. It is certain, that the atmosphere does not derive its heat from the immediate action of the solar rays. These rays, in traversing that subtle and transparent medium, are but slightly refracted, and, meeting with little obstruction, neither lose nor communicate much of their influence.

* Phil. Trans. 1777.

We are assured of this by many experiments; and we know, that air, in the focus of a burning glass, is never heated till some solid body be introduced The atmosphere, therefore, is warmed by the earth, from the surface of which a quantity of heat is continually flowing off, and ascending through the different strata of the air into the regions of vacuity, or of æther. But this ascent, on the whole, is uniform; because there is a certain temperature which, though varied by periodical vicissitudes, remains under every parallel the same, as to its mean quantity. Every stratum, therefore, of the atmosphere, whatever be its height, gives out, at a medium, the same quantity of heat that it receives; in other words, its mean temperature is constant, and neither increases nor decreases, on the whole.

3. Let there be three strata, then, of the atmosphere of the same thickness x, and contiguous to one another; so that, if x be the distance of the first from the surface of the earth, that of the second may be $x+\dot{x}$, and of the third $x+2x$. Let h, h', h'', be the heats of the strata, and Δ, Δ', Δ'', their densities respectively. Then, since the quantity of heat, communicated in an instant from one stratum of a fluid to a contiguous stratum, must be, as the difference of their temperatures, multiplied into the density of the colder, and divided by the density of the warmer, the heat communicated, in an instant, from the first stratum to the second, =

$(h-h')\frac{\Delta'}{\Delta}$; and that communicated by the second to the third, $=(h'-h'')\frac{\Delta''}{\Delta'}$. But, since the difference of Δ and Δ' is indefinitely small, as also that of Δ' and Δ'', we have $\frac{\Delta'}{\Delta}=1$, and $\frac{\Delta''}{\Delta'}=1$; so that the heat gained by the middle stratum is $=h-h'$, and that lost by it $=h'-h''$. Now, these two quantities must be equal, in order that the temperature of the stratum may remain uniform, that is, $h-h'=h'-h''$; or, in other words, the heat of the first stratum exceeds the heat of the second, as much as the heat of the second exceeds the heat of the third. Therefore, the heat of the successive strata must decrease, by equal differences, as we ascend through equal spaces, into the atmosphere; and, in general, the differences of temperature must be proportional to the differences of elevation.

It is to be understood, however, that this law is subject to certain anomalies, both annual and diurnal, and those intermixed with other accidental irregularities, which it would be difficult, perhaps impossible, to ascertain. All that can be said of it is, that it is the law which nature tends to observe, and that the sum of the deviations from it, on the one side, is probably equal to the sum of those on the other. In an effect that is perpetually subject to the action of accidental and unknown causes, the discovery of a mean, from which the departures on the opposite sides are equal, is all that we can reasonably expect; and it is sufficient for us to know, that, though any particular conclusion may involve an error, yet, if a multitude of instances be taken, the errors will certainly correct one another.

4. If, therefore, H be the heat at the surface of the earth, and h the heat at any given height a, above the surface, the heat, at any other height, as x, will be $H-\frac{(H-h)x}{a}$. At a medium, it is found, that Fahrenheit's thermometer falls a degree for every 300 feet that we ascend into the atmosphere; so that, if x is expressed in fathoms, the heat, at that height, is $=H-\frac{x}{50}$.

5. But though we are thus led to conclude, that the decrease of heat in the superior strata of the atmosphere is proportional to their elevation, there is no reason to suppose, that the condensation produced by that decrease is also uniform. Indeed, the experiments of General Roy have placed it beyond all doubt, that the variations in bulk of a given quantity of air are, by no means, proportional to its variations of temperature. Those experiments, though very numerous, are too few to ascertain exactly the law which connects these variations, and we must have recourse to reasoning, in order to supply this defect. Let us suppose that air of a given temperature, for instance, of 32°, by

the loss of one degree of heat, is contracted $\frac{1}{411}$, or the part m, of its whole bulk; its bulk, therefore, when of the temperature 31°, will be $1-m$. By the loss of another degree of heat, its temperature will be reduced to 30°, and its contraction will not be m, as before, but $m(1-m)$, which, subtracted from $1-m$, its bulk, when of the temperature 31°, will give its bulk when of the temperature 30°, $=1-2m+m^2=(1-m)^2$ In like manner, after the loss of 3° of heat, the bulk of the same given quantity of air is shown to be $(1-m)^3$; and, in general, its bulk is as that power of $1-m$, which is denoted by the difference between 32° and the given temperature. If, therefore, h be the heat of a given quantity of air, $(1-m)^{32-h}$ will be the space occupied by that air, supposing always that the compressing force is given.

6. This formula assigns a finite magnitude to the air as long as the diminution of its heat is less than infinite; for as $1-m$ is less than unity, when h becomes negative and infinite, $(1-m)^{32-h}$ becomes then, and not till then, $=0$. When h is affirmative, and greater than 32, $(1-m)^{32-h}$ becomes greater than 1, and increases continually, being infinite when h is infinite. When $32-h$ is not very great, then $(1-m)^{32-h}=1+(h-32)m$ nearly, which agrees with the hypothesis of uniform contraction and dilatation in moderate temperatures.

This formula also represents, with tolerable exactness, the experiments which General Roy made with the manometer, excepting in one circumstance; for the formula makes the expansion increase with the heat continually, though not uniformly; whereas the experiments give the greatest expansion between the temperatures of 60° and 70°. But this seems to be so anomalous a fact, that it looks more like some accidental effect, produced from the particular manner of making the experiments, than a part of that law of nature, which connects the variations of bulk in bodies with their variations of temperature.

7. But this is not the only irregularity to which the expansion of air by heat, and its contraction by cold, appear to be subject. We learn from the manometrical experiments of the same excellent observer, that a given variation of temperature is accompanied with more or less variation of bulk, according as the air is compressed by a greater or a less force. Air, for instance, compressed by the weight of an entire atmosphere, was expanded by the 180 degrees from freezing to boiling, no less than 484 of those parts, whereof, at the temperature 32°, it occupied 1000. But the same air, when compressed only by $\frac{1}{5}$ of an atmosphere, was, by the same difference of heat, expanded no more than 141 parts; and that though the heat of boiling water was applied to it for an hour together. It is not easy either to assign the cause, or to determine

the law of this inequality. General Roy has, indeed, constructed a table of the correction to be made on account of it; which proceeds on the supposition, that the expansion, for one degree of heat, decreases in the same proportion that the column of mercury in the barometer exceeds a given length. This given length is nearly $=4.5$ inches; so that if b be the length of the column of mercury in the barometer, and .00252 the expansion for one degree of heat, when the barometer is at 30 inches, and the temperature of the air 32°, then $\frac{b-4.5}{25.5} \times .00252$, will be the expansion of air of the same temperature, for the same change of heat, when the mercury in the barometer stands at the height b. But this formula cannot be just, otherwise air, compressed by no greater a force than that of 4.5 inches of mercury, would be incapable of dilatation by heat, or contraction by cold.

8. It will agree equally well with the experiments, and will involve no contradiction, even in the extreme cases, to suppose, that the expansion for a certain degree of heat is as a certain power of the compressing force. If this power be called μ, m being the expansion for 1 degree of heat, when the mercury in the barometer is of the height b, the expansion for any other height of the mercury, as β, will be $\frac{\beta^\mu}{b^\mu} m$; and combining this with the for-

mer formula for expansion, (§ 5,) we have the space which air occupies, as far as it depends on temperature, $=\left(1-\frac{m\beta^{\mu}}{b^{\mu}}\right)^{32-h}$. From a comparison of General Roy's experiments, * μ appears to be between $\frac{1}{2}$ and $\frac{1}{3}$; and it must be confessed, that it is very difficult to assign its value within nearer limits. The form of the correction, however, if not its absolute quantity, may be found from what is here determined. The last of these must be ascertained by future experiments.

9. These inequalities belong to the temperature of the air; there is another that depends wholly on the compression. In deducing the rule for the measurement of heights by the barometer, it has hitherto been supposed, agreeably to the experiments of Mr Boyle and M. Mariotte, that the density of the air, while its temperature remains the same, is exactly as the force that compresses it. But the experiments referred to were not accurate enough to establish this law with absolute precision; and they left room to suspect a deviation from it, either when the compressing force is very great or very small. Accordingly, from experiments described in the ninth volume of the Mem. of Berlin, it appears that the elasticity of air of the temperature 55°, or the compressing force, in-

* Tab. 2 and 3, p. 701, 703, Trans. 1777. Part II.

creases more slowly than the density; so that, if the compressing force be doubled, the density will exceed the double by about a tenth part, &c. The law of this variation is expressed with tolerable exactness, by supposing, that if D be the density of the air, and F the force compressing it, then $D=F^{1+n}$, n being a very small fraction, nearly .0015.

10. It must be acknowledged, that new experiments are necessary to ascertain the law of this inequality with precision. But as the formula $D=F^{1+n}$ is very general, and might be rendered still more so, without affecting the method of integration that is to be employed, the result of that integration may be useful when our physical knowledge becomes more accurate. In the meantime, it may not be improper to remark, that the precise knowledge of the law which connects the compressing force with the density of elastic fluids, is an object well deserving the attention of natural philosophers. The determination of that law may go far to decide the question, whether the particles of such fluids are in contact or not; that is, whether the elasticity of each particle be a force that extends beyond the nearest particles, like the forces of magnetism and gravitation; or one which, like that of a spring, extends only to the bodies which are next it. It is an inquiry, therefore, of no less import-

ance in general physics than in that particular subject which we have here undertaken to examine.

11. There is one other correction to be applied to the height of a mountain, as it is usually found from observations of the barometer. This arises from the diminution of gravity, whether we ascend or descend from the surface of the earth. The effect of that diminution is to produce a twofold error; because, on the supposition of uniform gravity, the weight of each particle of air is computed too great, and the weight of the column of mercury in the barometer, that is not on the surface, is also reckoned too great. The effect of both these errors is of the same kind, tending to make the height less than it is in reality; yet it is only the first of them, and that too the least considerable, which has hitherto been taken into account.

12. It were to be wished, that, to the causes here enumerated, and that are to be introduced into the computation, we could add the operation of moisture, in altering the weight and elasticity of the air. But the law of that operation has not yet been discovered; and it will be sufficient to point out, in the conclusion of this paper, a method by which it may be determined from observations of the barometer itself.

Before proceeding to the investigation of the effect which all these inequalities together must produce, it is proper to remark, that the two inequalities in the expansion of air, taken notice of, (§ 5 and 7,) after having been discovered by General Roy, were applied by him to correct the height of mountains, measured by the barometer; but that it is by no means certain that he has given to those corrections the precise form which they ought to have. This, indeed, cannot be known, unless the effect of each inequality, on a single stratum, be first introduced into the differential equation between the density of the air and the height above the surface, and the amount of its effect on a whole column of air be deduced from thence by integration.

13. Let y, then, be the density of the air, at any height x above the surface of the earth, the heat at the surface being $=$H, expressed in degrees of Fahrenheit's thermometer. If also λ be such a number, that λx gives the degrees by which the thermometer stands lower at the height x than at the surface, (§ 4,) the temperature at the height x will be $=\mathrm{H}-\lambda x$; and, if the expansion of a given quantity of air, which occupies the space 1, and is of the temperature 32°, for 1° of heat, be called m, then, abstracting at present from that inequality of expansion which depends on pressure, we have the space occupied by that same quantity of air, when it is of the temperature $\mathrm{H}-\lambda x$, equal to $(1-m)^{32-\mathrm{H}+\lambda x}$: Or, making $32-\mathrm{H}=\tau$, we have the required space $=(1-m)^{\tau+\lambda x}$.

Now, if the given quantity of air, of which the bulk has been supposed $=1$, and the temperature $=32°$, be compressed by a column of air of the same density and temperature with itself, but of the height p, and if its density, in this case, be also called 1 ; then, in the case of its having any other temperature, as $H-\lambda x$, and being compressed by any other force, as $-\int y\dot{x}$, or the weight of the superincumbent air at the height x, we have

$$1 : y :: p : \frac{-\int yx}{(1-m)^{\tau+\lambda x}}, \text{ and likewise } y = \frac{-\int y\dot{x}}{p(1-m)^{\tau+\lambda x}}.$$

No account is here taken of the diminution of gravity, any more than of the departure of the law of the elasticity of air from direct proportionality to the density, (§ 8,) because it is convenient to consider the problem at first under the more simple view, where only the two first inequalities are introduced.

13. Since $y = \frac{-\int yx}{p(1-m)^{\ +\lambda x}}$ we have

$$py(1-m)^{\tau+\lambda x} = -\int y\dot{x}, \text{ and}$$

$$p\dot{y}(1-m)^{\tau+\lambda x} + p\lambda y\,(\log.\,1-m)\,(1-m)^{\tau+\lambda x}\dot{x} = -y\dot{x},$$

Or, $\frac{p\dot{y}}{y} + p\lambda\log.(1-m)x = -\frac{\dot{x}}{(1-m)^{\tau+\lambda x}}$.

Hence making $\log.(1-m)=g$, $\frac{p\dot{y}}{y} = -p\lambda g x - \frac{\dot{x}}{(1-m)^{\tau+\lambda x}}$, and $p\log.y + p\log.C = -p\lambda g x + \frac{1}{\lambda g\,(1-m)^{\tau+\lambda x}}$.

If D denote the density of the air at the surface of the earth, D will be the value of y, when $x=0$, and so $p(\log.D+\log.C)=\frac{1}{\lambda g(1-m)^{\tau}}$. Therefore $p\log.C=\frac{1}{\lambda g(1-m)^{\tau}}-p\log.D$; and so by substituting for $p\log.C$, $p(\log.y-\log.D)+\frac{1}{\lambda g(1-m)^{\tau}}=-p\lambda gx+\frac{1}{\lambda g(1-m)^{\tau+\lambda x}}$; or changing the signs, $p(\log.D-\log.y)-\frac{1}{\lambda g(1-m)^{\tau}}=p\lambda gx-\frac{1}{\lambda g(1-m)^{\tau+\lambda x}}$.

This equation exhibits, in general, the relation between the density of any stratum of air, and the height of that stratum above the surface of the earth, on the suppositions that the heat of the atmosphere decreases uniformly as we ascend, and that the contraction produced in air by cold, observes the law described in § 5. It might be considered as an equation to a curve, of which the abscissæ represented the height of the different strata of the atmosphere, and the ordinates, the densities of those strata: this curve would evidently be different from the logarithmic, but would be found to have certain relations to it not uninteresting, and not difficult to trace, if we had leisure for such a digression.

14. Let us now suppose that z is the whole height to be measured, and that Δ is the density at

that height, the temperature there being also found $=h$, by observation. If then x become $=z$, and $y=\Delta$, we will also have $\lambda z=H-h$, and $\tau+\lambda z=32-H+H-h=32-h=r-h$, making $r=32$. Also $\lambda=\frac{H-h}{z}$. Therefore, by substituting these values of y, x, λ, and $\tau+\lambda z$, in the preceding equation, we have, $p(\log.D-\log.\Delta)-\frac{z}{g(H-h)(1-m)^{\tau}}=pg(H-h)-\frac{z}{g(H-h)(1-m)^{r-h}}$. Hence, by transposition, &c. $gp(H-h)(\log.D-\log.\Delta-(H-h)g)=z\left((1-m)^{H-r}-(1-m)^{h-r}\right)$; and

$$x=\frac{gp(H-h)(\log.D-\log.\Delta-(H-h)g)}{(1-m)^{H-r}-(1-m)^{h-r}}.$$

Thus the height of any column of air is expressed in terms of the density, and of the temperature at the top and bottom of it; the equation for the height, though an exponential one in its general form, admitting of an easy resolution, from the circumstance of λz being given by the observations of the thermometer.

15. That this formula may be applied to the measurement of heights, it is necessary to introduce into it the lengths of the columns of mercury in the barometer, instead of the densities of the air, at the lower and upper stations. Let b be the height at which the mercury stands in the lower barometer,

and β that at which it stands in the higher barometer; then, since b is the compressing force at the surface of the earth, we have

$D = \frac{b}{(1-m)^{r-H}}$; and, for a like reason, $\Delta = \frac{\beta}{(1-m)^{r-h}}$. Therefore, $\log. D = \log. b - (r-H)g$, and $-\log.\Delta = -\log.\beta + (r-h)g$. Hence $\log. D - \log.\Delta = \log. b - \log.\beta + (H-h)g$, and substituting for $\log. D - \log.\Delta$ in the formula of the last section,

$$z = \frac{gp(H-h)(\log.b - \log.\beta)}{(1-m)^{H-r} - (1-m)^{h-r}}.$$

16. This is the exact value of z, or of the whole height to be measured, on the supposition that the heat of the atmosphere decreases uniformly as the height increases; and that the contraction for a given difference of heat decreases according to the law described in § 5. But, in order that it may be more convenient for computation, and may be more easily compared with the formula now in use, the quantity $\frac{1}{(1-m)^{H-r} - (1-m)^{h-r}}$ must be reduced into a series. Now $\frac{1}{(1-m)^{H-r} - (1-m)^{h-r}} = \frac{(1-m)^r}{(1-m)^H - (1-m)^h}$. But from the nature of logarithms, (g being, as before, the logarithm of $1-m$) $(1-m)^H = 1 + Hg + \frac{Hg^2}{2} + \frac{H^3g^3}{6} + \&c.$ And

$-(1-m)^h=-1-hg-\frac{h^2g^2}{2}-\frac{h^3g^3}{6}-$ &c. Therefore

$$\frac{(1-m)^r}{(1-m)^{\mathrm{H}}-(1-m)^h}=\frac{1+rg+\frac{r^2}{2}g^2+\frac{r^3}{6}g^3+\&c.}{(\mathrm{H}-h)g+\frac{\mathrm{H}^2-h^2}{2}g^2+\frac{\mathrm{H}^3-h^3}{6}g^3+\&c.};$$

and $$\frac{g(\mathrm{H}-h)(1-m)^r}{(1-m)^{\mathrm{H}}-(1-m)^h}=\frac{1+rg+\frac{r^2}{2}g^2+\frac{r^3}{6}g^3+\&c.}{1+\frac{\mathrm{H}+h}{2}g+\frac{\mathrm{H}^2+\mathrm{H}h+h^2}{6}g^2+\&c.}.$$

Hence $$z=p(\log.b-\log.\beta)\left(\frac{1+rg+\frac{r^2}{2}g^2+\frac{r^3}{6}g^3+\&c.}{1+\frac{\mathrm{H}+h}{2}g+\frac{\mathrm{H}^2+\mathrm{H}h+h^2}{6}g^2+\&c.}\right).$$

17. These series will not converge fast, unless rg, $\mathrm{H}g$, and hg, be all of them quantities much less than unity. Now, as m, or the expansion of air of the temperature r, for $1°$ of heat, is, in fact, very small, being nearly $=.00245$, and as g, or the logarithm of $1-m$, must, of consequence, be nearly $=-m=-.00245$, it is plain, that, in all moderate temperatures, these series will converge with great rapidity; though, in extreme cases, where z is supposed vastly great, and where h may be negative, and also great, the series in the denominator may converge so slowly that recourse must be had to the formula in § 15, from which no quantities are rejected.

When m, and, of consequence, g, are very small, and when H and h do not differ much from r, the preceding formula, agreeably to a remark in § 6, will comprehend the case of uniform expansion, and will give the same expression for the height, that would be derived from considering only the equable decrease of heat as we ascend in the atmosphere. Now, as in the case supposed, we may reject all the powers of g but the first, and may also suppose $g=-m$, we have

$$z=p(\log.b-\log.\beta)\left(\frac{1-rm}{1-\frac{H+h}{2}m}\right), \text{ or}$$

$$z=p\left(1+\left(\frac{H+h}{2}-r\right)m\right)(\log.b-\log.\beta).$$

18. This last is precisely the formula of M. Deluc, if we give to p, r, and m, the proper values.* It was discovered by that ingenious and indefatigable observer, without any inquiry into the propagation of heat through the atmosphere, the principle on which it depends; and, that so near an approximation to the truth should have been thus obtained, is to be considered as a singular instance of sagacity or of good fortune. For if the heat of the

* If we take M. Deluc's rule, as improved by the later observations of General Roy and Sir George Shuckburgh, $p=4342.9448=$ the modulus of the tabular logarithms multiplied by 10000: $r=32°$ and $m=.00245$ nearly. It is unnecessary to remark, that the logarithms understood in all these formulas are hyperbolic logarithms, and that the multiplication of them by p is saved, by using the tabular logarithms, and making the first four places of them, excluding the index, integers.

air diminished, not in the simple ratio of the increase of the height, but in that of any power of it, so as to be expressed by $H-\lambda x^n$, then, by computing as has been done above, we should find $z = p\left(1+m\left(\frac{nH+h}{n+1}-r\right)\right)\log.\frac{b}{\beta}$. Here the temperature from which r, or the fixed temperature, is to be subtracted, is not $\frac{H+h}{2}$, but $\frac{nH+h}{n+1}$; and this is a formula which conjecture or experiment alone would scarcely have discovered.

It is farther to be remarked of the formula $z = p\left(1+m\left(\frac{H+h}{2}-r\right)\right)\log.\frac{b}{\beta}$, that it is rigorously just, if we suppose the temperature $\frac{H+h}{2}$ to be uniformly diffused through the column of air, of which the height is to be measured, as is done by Dr Horsley in his theory of M. Deluc's rules; * but that, on a supposition, more conformable to nature, of the heat diminishing in the same proportion as the height increases, it is only an approximation to the truth, or the first term of a series, whereof the other terms are rejected as inconsiderable.

19. The amount of the terms, which are thus rejected, comes now to be considered; and it will be ascertained with sufficient accuracy, if we com-

---

* Phil. Trans. Vol. LXIV. Part 1.

pute the second term of the series, or that which involves in it m^2. Now,

$$\frac{1+rg+\frac{r^2}{2}g^2+ \&c.}{1+\frac{H+h}{2}g+\frac{H^2+Hh+h^2}{6}g^2+ \&c.}=$$

$$1+\left(r-\frac{H+h}{2}\right)g+\left(\frac{r^2-r(H+h)}{2}+\frac{H^2+4Hh+h^2}{12}\right)g^2;$$

$$\text{and } g=\log.(1-m)=-m+\frac{m^2}{2}- \&c.$$

so that $g^2=$ m^2- &c.

Therefore, by substitution, $\dfrac{1+rg+\frac{r^2}{2}g^2}{1+\frac{H+h}{2}g+\frac{H^2+Hh+h^2}{6}g^2}=$

$$1+\left(\frac{H+h}{2}-r\right)m+\left(\frac{r}{2}-\frac{H+h}{4}+\frac{r^2-r(H+h)}{2}+\right.$$

$$\left.\frac{H^2+4Hh+h^2}{12}\right)m^2.$$

This is the coefficient of $p \log.\frac{b}{\beta}$, which gives z, corrected both for the temperature of the air and the first inequality of expansion, (§ 5.) The term $\left(\frac{H+h}{2}-r\right)m$, is M. Deluc's correction, as has been already observed, the third term, viz. $\left(\frac{r}{2}-\frac{H+h}{4}+\frac{r^2-r(H+h)}{2}+\frac{H^2+4Hh+h^2}{12}\right)m^2$, contains not only a part which depends on the equable decrease of heat as we ascend in the atmosphere, but also one which arises from the above mentioned inequality of expansion.

20. The term involving m^2, that has now been computed, will rarely amount to any thing considerable. The coefficient of it vanishes when both H and h are equal to r, but increases as these two quantities recede from r on either side. In no instance where the barometer is to be applied to actual measurement, will the correction probably be found greater than in determining the height of Coracon above the level of the South Sea, where H, or the height of the thermometer at that level, was $84\frac{1}{2}°$, and h, or the height of the thermometer at the top of the mountain, $43\frac{1}{2}°$; the coefficient of m^2 comes out, in this case $+426$, and m^2 being $=.000006=(.00245)^2$, the correction $=.00259$, or nearly $\frac{1}{400}$ of the height of the mountain, as found before any correction was applied, or $=40$ feet nearly. It is to be remarked, too, that, for every value of H, or of the temperature at the lower station, there are two values of h, or the temperature at the upper station, that make the coefficient, $\frac{r}{2}-\frac{H+h}{4}+\frac{r(r-H-h)}{2}+\frac{H^2+4Hh+h^2}{12}$, and, of consequence, the correction depending on it equal to nothing. This is evident from the nature of the coefficient; but, as the law by which this last increases and decreases is by no means simple, it were convenient to have it reduced into a table, for the different values that might be assigned to H and h,

from which it would be immediately obvious in what cases it was to be taken into account, and when it might safely be omitted.

But though this correction may sometimes be of consequence enough to be included in the measurement of heights, it is certain that it may be safely neglected in the computation of the other corrections. For the error thereby committed in the estimation of a new correction, will be nearly the same part of the former correction, that the new one is of the whole height. If, for instance, the new correction be $\frac{1}{100}$ of the whole height, the error committed in estimating it will be but $\frac{1}{100}$ of the former correction; and, if that did not exceed $\frac{1}{400}$, the error in question will not exceed $\frac{1}{40000}$ of the whole height.

21. In computing the effect of the second inequality of expansion, described § 8, we may, therefore, abstract from the last inequality, and may even suppose, with M. Deluc, that the temperature, which is a mean between those of the extremities of a column of air, is uniformly diffused through that column. Let the excess of that mean, above the temperature r, or $\frac{H+h}{2}-r=f$; and let β, the height of the mercury in the uppermost baro-

meter, be considered as variable. Then taking the formula of § 8, and supposing m to be the expansion for 1° of heat, when the mercury in the barometer is of a given height, which we shall here call γ,* (to avoid the confusion that would arise from naming it, as in the art. above referred to,) and retaining all the other denominations as before, we have $y = \frac{-f y \dot{x}}{p\left(1 + \frac{fm}{\gamma^{\mu}}\beta^{\mu}\right)}$. Hence

$py\left(1 + \frac{fm\beta^{\mu}}{\gamma^{\mu}}\right) = -f y \dot{x}$, so that, taking the fluxions,

$p\dot{y} + \frac{pfm\beta^{\mu}\dot{y}}{\gamma^{\mu}} + \frac{pfm\mu y\beta^{\mu-1}\dot{\beta}}{\gamma^{\mu}} = -y\dot{x}$ and, dividing by y,

$= -\frac{p\dot{y}}{y} - \frac{fmp\beta^{\mu}\dot{y}}{\gamma^{\mu}y} - \frac{\mu fmp\beta^{\mu-1}\dot{\beta}}{\gamma^{\mu}}$. To exterminate from this equation y and $\dot{y}$, it is to be remarked, that

---

* According to the experiments of General Roy, above quoted, the expansion of air for 1° of heat, at the temperature 32°, is .00245 nearly, that air being compressed at the same time by the weight of a column of mercury 29.5 inches high. As we have supposed m, in the preceding computations, to be .00245, we must suppose $\gamma = 29.5$. The formula supposed here to give the space occupied by the air, so far as heat is concerned, viz. $1 + \frac{fm}{\gamma^{\mu}}\beta^{\mu}$, is changed from the exponential expression of § 8, in consequence of what has been just observed about the effect of neglecting one inequality in the computation of another.

$y=\frac{\beta}{p\left(1+\frac{fm}{\gamma^\mu}\beta^\mu\right)}$, and that therefore

$\frac{\dot{y}}{y}=\frac{\dot{\beta}}{\beta}-\frac{\mu fm\beta^{\mu-1}\dot{\beta}}{\gamma^\mu+fm\beta^\mu}$. Hence, by substitution, $\dot{x}=$

$$p\left(-\frac{\dot{\beta}}{\beta}+\frac{\mu fm\beta^{\mu-1}\dot{\beta}}{\gamma^\mu+fm\beta^\mu}-\frac{(1+\mu)fm\beta^{\mu-1}\dot{\beta}}{\gamma^\mu}+\frac{\mu f^2m^2\beta^{2\mu-1}\dot{\beta}}{\gamma^\mu(\gamma^\mu+fm\beta^\mu)}\right).$$

But $\frac{\mu f^2m^2\beta^{2\mu-1}\dot{\beta}}{\gamma^\mu(\gamma^\mu+fm\beta^\mu)}=\frac{\mu fm\beta^{\mu-1}\dot{\beta}}{\gamma^\mu}-\frac{\mu fm\beta^{\mu-1}\dot{\beta}}{\gamma^\mu+fm\beta^\mu}$;

therefore $\dot{x}=p\left(-\frac{\dot{\beta}}{\beta}-\frac{fm\beta^{\mu-1}\dot{\beta}}{\gamma^\mu}\right)$, the other terms destroying one another. By integration, then, $x=p(-\log.\beta-\frac{fm\beta^\mu}{\mu\gamma^\mu}+C)$. If C be taken such that x may vanish when $\beta=b$, the height of the mercury in the lower barometer, we will have

$$x=p\left(\log.\frac{b}{\beta}+\frac{fm(b^\mu-\beta^\mu)}{\mu\gamma^\mu}\right).$$

22. That it may appear wherein this formula differs from the ordinary one, instead of b^μ and β^μ, we must introduce log.b, and log.β, which, when b and β are not very unequal, may be done without difficulty. For we have

$$\frac{b^\mu}{\gamma^\mu}=1+\mu\log.\frac{b}{\gamma}+\frac{\mu^2}{2}\left(\log.\frac{b}{\gamma}\right)^2+\frac{\mu^3}{6}\left(\log.\frac{b}{\gamma}\right)^3+\text{\&c.};$$

also $\frac{\beta^\mu}{\gamma^\mu}=1+\mu\log.\frac{\beta}{\gamma}+\frac{\mu^2}{2}\left(\log.\frac{\beta}{\gamma}\right)^2+\frac{\mu^3}{6}\left(\log.\frac{\beta}{\gamma}\right)^3+\text{\&c.}$

Therefore $\frac{b^{\mu}-\beta^{\mu}}{\gamma^{\mu}}=\mu\left(\log.\frac{b}{\gamma}-\log.\frac{\beta}{\gamma}\right)+\frac{\mu^2}{2}\left(\left(\log.\frac{b}{\gamma}\right)^2\right.$ $\left.-\left(\log.\frac{\beta}{\gamma}\right)^2\right)+\&c.$ That is, $\frac{b^{\mu}-\beta^{\mu}}{\gamma^{\mu}}=\mu\log.\frac{b}{\beta}+$ $\frac{\mu^2}{2}\log.\frac{b\beta}{\gamma^2}\times\log.\frac{b}{\beta}$, rejecting all the terms which involve powers, of $\log.\frac{b}{\gamma}$, of $\log.\frac{\beta}{\gamma}$, and of μ, higher than the square. Hence also,

$$\frac{fm\,(b^{\mu}-\beta^{\mu})}{\mu\gamma^{\mu}}=fm\log.\frac{b}{\beta}+\frac{\mu fm}{2}\log.\frac{b\beta}{\gamma^2}\times\log.\frac{b}{\beta},\text{ and}$$

$$x=p\left(\log.\frac{b}{\beta}+\frac{fm(b^{\mu}-\beta^{\mu})}{\mu\gamma^{\mu}}\right)=$$

$$p\left(\log.\frac{b}{\beta}+fm\log.\frac{b}{\beta}+\frac{\mu fm}{2}\log.\frac{b\beta}{\gamma^2}\times\log.\frac{b}{\beta}\right);\text{ or}$$

$$x=p\log.\frac{b}{\beta}\left(1+fm+\frac{\mu fm}{2}\log.\frac{b\beta}{\gamma^2}\right).$$

23. This formula includes the correction to be made for that inequality of the expansion of air by heat which depends on its compression, and which was described at the 7th and 8th articles. The first term of the formula, viz. $p\log.\frac{b}{\beta}$, is the difference of the tabular logarithms of b and β. The second, viz. $fmp\log.\frac{b}{\beta}$, is M. Deluc's correction, and the same that was already investigated, § 17. The third, viz. $\frac{\mu fm}{2}\log.\frac{b\beta}{\gamma^2}\times p\log.\frac{b}{\beta}$ is the correction

for the above mentioned inequality of expansion. It is of a form very convenient for computation; for the former correction being $=fmp\log.\frac{b}{\beta}$, we need only multiply it by $-\frac{\mu}{2}\log.\frac{b\beta}{\gamma^2}$ to have the third term of the formula, or the correction required. It must be remembered, that $\log.\frac{b\beta}{\gamma^2}$ signifies the hyperbolic logarithm of $\frac{b\beta}{\gamma^2}$.

The exact amount of this correction cannot be known, till μ be defined by experiments on the expansibility of air under different degrees of compression; those which General Roy has made, though excellent, not being perfectly sufficient for that purpose. If we suppose $\mu=\frac{1}{2}$, and if, as an example, we take $b=29$ inches, and $\beta=24$, γ being $=29.5$, then we will find $\log.\frac{b\beta}{\gamma^2}=-.22$ nearly, which, multiplied into $\frac{\mu}{2}$, or into $\frac{1}{4}$, is $-\frac{1}{16}$ nearly, and this multiplied into M. Deluc's correction, gives the correction for the compression. The former is, therefore, to be diminished by $\frac{1}{16}$, before it be applied to the difference of the tabular logarithms, to give the true height of the one barometer above the other. In other cases, the proportional part, to be added or subtracted, will be greater as β becomes less, or as the height becomes greater: It will be $=0$, when $b\beta=\gamma^2$; affirmative, when $b\beta$ is greater than γ^2; and negative when it is less.

24. There remain to be considered the two corrections that depend, one, on the relation between the density of the air and the force compressing it; the other, on the diminution of gravity as we ascend from the surface of the earth. It was observed, (§ 9,) that, if D denote the density of the air, and F the compressing force, $D=F^{1+n}$. But the force compressing a stratum of the atmosphere at the height x above the surface of the earth, and of the density y, which, on the supposition of uniform gravity, is denoted by $-\int y\dot{x}$, on that of gravity decreasing as the ν power of the distance from the centre of the earth, is denoted by $-\int\frac{s^\nu}{(s+x)^\nu}y\dot{x}$; where s is the semidiameter of the earth. This is evident, because the weight of each stratum of air is proportional to its density, multiplied into the accelerating force which draws the particles of it toward the earth. Now, let q be the length of such a column of mercury, that air, compressed by it, would be of the same density with the mercury itself, which density, in all the preceding investigations, is understood to be constant, and to be $=1$;*

* The mercury in the barometers is supposed to be reduced to a fixed temperature, by the application of a correc-

then, $1 : y :: q^{1+n} : \dfrac{\left(-\int \frac{s^v}{(s+x)^v} y\dot{x}\right)^{1+n}}{(1-m)^{\tau+\lambda x}}$, and

$$y = \frac{\left(-\int \frac{s^v}{(s+x)^v} y\dot{x}\right)^{1+n}}{q^{1+n}(1-m)^{\tau+\lambda x}}, \text{ or } y^{\frac{1}{1+n}} = \frac{-\int \frac{s^v}{(s+x)^v} y\dot{x}}{q(1-m)^{\frac{\tau+\lambda x}{1+n}}}.$$

In which formula, all the inequalities that have been enumerated are expressed, except that which was considered in the two preceding articles. Hence, multiplying by $q(r-m)^{\tau+\lambda x}$, and taking the fluxions, there comes out,

$$q\left(\frac{1}{1+n} y^{\frac{1}{1+n}-1} \dot{y}(1-m)^{\frac{\tau+\lambda x}{1+n}} + \frac{\lambda g}{1+n} y^{\frac{1}{1+n}} (1-m)^{\frac{\tau+\lambda x}{1+n}} \dot{x}\right)$$

$$= -\frac{s^v y\dot{x}}{(s+x)^v}.$$

Dividing therefore by y,

$$q\left(\frac{1}{1+n} y^{\frac{1}{1+n}-2} \dot{y}(1-m)^{\frac{\tau+\lambda x}{1+n}} + \frac{\lambda g}{1+n} y^{\frac{1}{1+n}-1} (1-m)^{\frac{\tau+\lambda x}{1+n}} \dot{x}\right)$$

$= -\dfrac{s^v \dot{x}}{(s+x)^v}$; and making $y^{\frac{1}{1+n}-1} = v$, and, conse-

---

tion on account of the thermometers attached to them, after the manner of M. Deluc, or of General Roy; the latter reduces the mercury always to the temperature of 32°. When the difference of temperature is not very great in the two barometers, the correction of their heights may be made according to the very ingenious remark of the astronomer royal. Phil. Trans. Vol. LXIV. Part I. p. 164.

quently, $\frac{-n}{1+n}y^{\frac{1}{1+n}-2}\dot{y}=\dot{v}$, we have

$$\dot{v}(1-m)^{\frac{\tau+\lambda x}{1+n}}-\frac{n\lambda g}{1+n}v(1-m)^{\frac{\tau+\lambda x}{1+n}}\dot{x}=\frac{ns^{\nu}\dot{x}}{q(s+x)^{\nu}}.$$ This equation will become integrable if it be multiplied by $(1-m)^{-\lambda x}$, for it is then

$$\dot{v}(1-m)^{\frac{\tau-n\lambda x}{1+n}}-\frac{n\lambda g}{1+n}v(1-m)^{\frac{\tau-n\lambda x}{1+n}}\dot{x}=\frac{ns^{\nu}\dot{x}}{q(s+x)^{\nu}(1-m)^{\lambda x}};$$

and so $v(1-m)^{\frac{\tau-n\lambda x}{1+n}}+\mathrm{C}=\frac{ns^{\nu}}{q}\int\frac{\dot{x}}{(s+x)^{\nu}(1-m)^{\lambda x}}$.

But $v=y^{\frac{1}{1+n}-1}=y^{-\frac{n}{1+n}}$, therefore,

$$y^{-\frac{n}{1+n}}(1-m)^{\frac{\tau-n\lambda x}{1+n}}+\mathrm{C}=\frac{ns^{\nu}}{q}\int\frac{\dot{x}}{(s+x)^{\nu}(1-m)^{\lambda x}}.$$

25. It is necessary to introduce β into this formula, by substituting for y, its value,

$$=\frac{\left(\frac{s^{\nu}}{(s+x)^{\nu}}\beta\right)^{1+n}}{q^{1+n}(1-m)^{\tau+\lambda x}};$$ and, therefore, as

$$y^{\frac{n}{1+n}}=\frac{s^{n\nu}\beta^{n}}{q^{n}(s+x)^{n\nu}(1-m)^{\frac{n(\tau+\lambda x)}{1+n}}},$$ we have

$$\frac{q^{n}(s+x)^{\nu n}(1-m)^{\frac{n(\tau+\lambda x)}{1+n}}}{s^{\nu n}\beta^{n}}(1-m)^{\frac{\tau-n\lambda x}{1+n}}+\mathrm{C}=$$

$$\frac{ns^{\nu}}{q}\int\frac{\dot{x}}{(s+x)^{\nu}(1-m)^{\lambda x}}, \text{ or, } \frac{q^{n}(s+x)^{\nu n}(1-m)^{\tau}}{s^{\nu n}\beta^{n}}+C=$$

$$\frac{ns^{\nu}}{q}\int\frac{\dot{x}}{(s+x)^{\nu}(1-m)^{\lambda x}}.$$

26. In the cases which actually take place in nature, ν is either equal to $+2$, or to -1. It is equal to $+2$, when the barometer is raised above the surface of the earth, and to -1, when it is depressed below it. When $\nu=+2$, the last equation becomes

$$\frac{q^{n}(s+x)^{2n}(1-m)^{\tau}}{s^{2n}\beta^{n}}+C=\frac{ns^{2}}{q}\int\frac{\dot{x}}{(s+x)^{2}(1-m)^{\lambda x}}.$$

When x is supposed very small in comparison of s, the fluent $\int\frac{\dot{x}}{(s+x)^{2}(1-m)^{\lambda x}}$ may be expressed by a series, converging with such rapidity, that the two first terms will be sufficient for the present purpose. Now, as $\frac{1}{(s+x)^{2}}=\frac{1}{s^{2}\left(1+\frac{x}{s}\right)^{2}}=\frac{1}{s^{2}}\left(1-\frac{2x}{s}\right)$ nearly,

$$\frac{ns^{2}}{q}\int\frac{\dot{x}}{(s+x)^{2}(1-m)^{\lambda x}} \text{ becomes } =$$

$$\frac{n}{q}\int\overline{(1-m)^{-\lambda x}\dot{x}\left(1-\frac{2x}{s}\right)}=\frac{n}{q(1-m)^{\lambda x}}\left(-\frac{1}{\lambda g}+\frac{2x}{\lambda gs}+\frac{2}{\lambda^{2}g^{2}s}\right).$$

Therefore, $\frac{q^{n}(s+x)^{2n}(1-m)^{\tau}}{s^{2n}\beta^{n}}+C=$ $\frac{n}{q(1-m)^{\lambda x}}\left(-\frac{1}{\lambda g}+\frac{2x}{\lambda gs}+\frac{2}{\lambda^{2}g^{2}s}\right)$. To define C, x must be put $=0$, and $\beta=b$, so that

$\frac{q^n(1-m)^\tau}{b^n}+C=\frac{n}{q}\left(-\frac{1}{\lambda g}+\frac{2}{\lambda^2 g^2 s}\right)$, and $C=$

$-\frac{q^n(1-m)^\tau}{b^n}+\frac{n}{q}\left(-\frac{1}{\lambda g}+\frac{1}{\lambda^2 g^2 s}\right)$. If this value be substituted for C, and if all the terms be divided by $(1-m)^\tau$, we shall have

$$\frac{q^n(s+x)^{2n}}{s^{2n}\beta^n}-\frac{q^n}{b^n}=\frac{n}{q\lambda g}\left(\frac{1}{(1-m)^\tau}-\frac{1}{(1-m)^{\tau+\lambda x}}+\frac{2x}{s(1-m)^{\tau+\lambda x}}-\frac{2}{\lambda g s(1-m)^\tau}+\frac{2}{\lambda g s(1-m)^{\tau+\lambda x}}\right).$$

The approximation which has been used here for finding the fluent $\int\frac{\dot{x}}{(s+x)^2(1-m)^{\lambda x}}$, was sufficiently exact, because no terms have been rejected but such as are divided by s^2, and which, of consequence, are extremely small in respect of the rest.

27. We are now to suppose, that x becomes equal to z, or to the whole height that is to be measured; then also, $\tau+\lambda x=r-h$, $\lambda=\frac{H-h}{z}$, and $\tau=r-H$, as in § 14; and so by substitution,

$$\frac{q^n(s+z)^{2n}}{s^{2n}\beta^n}-\frac{q^n}{b^n}=$$

$$\frac{nz}{qg(H-h)(1-m)^{r-H}}\left(1-\frac{1}{(1-m)^{H-h}}-\frac{2z}{gs(H-h)}+\frac{2z}{s(1-m)^{H-h}}+\frac{2z}{gs(H-h)(1-m)^{H-h}}\right)=\frac{nz}{qg(H-h)}$$

$$\left((1-m)^{H-r}-(1-m)^{h-r}-\frac{2z}{g^s(H-h)}(1-m)^{H-r}+\right.$$

$$\left.\frac{2z}{s}(1-m)^{h-r}+\frac{2z}{gs(H-h)}(1-m)^{h-r}\right).$$

The value of z is to be found from this equation; and as the first step in the approximation, we may suppose s so great in respect of z, that $s+z=s$, nearly; and, also, that all the terms divided by s vanish; which, in fact, is the same thing with supposing the force of gravity to be uniform. We have, then,

$$\frac{q^n}{\beta^n}-\frac{q^n}{b^n}=\frac{nz}{qg(H-h)}\left((1-m)^{H-r}-(1-m)^{h-r}\right), \text{ or,}$$

$$z=\frac{\frac{1}{n}q^{1+n}g(H-h)\left(\frac{1}{\beta^n}-\frac{1}{b^n}\right)}{(1-m)^{H-r}-(1-m)^{h-r}}.$$

28. This is the exact value of z, on the supposition that gravity is uniform, and that the elasticity of the air is not simply as its density, but as the power of it denoted by $\frac{1}{1+n}$. But if we content ourselves with an approximation, which the smallness of n renders easy, the logarithms of b and β may be introduced, and the formula will become similar to that which was formerly investigated. For $\frac{1}{\beta^n}$, or $\beta^{-n}=1-n\log.\beta+\frac{n^2}{2}(\log.b)^2-$

$\frac{n^3}{6}(\log.b)^3+$ &c. When n is very small, as in the present case, this series converges with extreme rapidity; and the terms involving n^3, &c. may safely be rejected. Therefore,

$$\frac{1}{\beta^n}-\frac{1}{b^n}=1-n\log.\beta+\frac{n^2}{2}(\log.\beta)^2-1+n\log.b-\frac{n^2}{2}(\log.b)^2=$$

$$n(\log.b-\log.\beta)-\frac{n^2}{2}\left((\log.b)^2-(\log.\beta)^2\right).$$ Hence,

$$z=\frac{q^{1+n}g(\mathrm{H}-h)\left(\log.b-\log.\beta-\frac{n}{2}(\log.b)^2+\frac{n}{2}(\log.\beta)^2\right)}{(1-m)^{\mathrm{H}-r}-(1-m)^{h-r}}.$$

29. When n vanishes altogether, the value of z, assigned by this formula, coincides, as it ought to do, with that which was investigated, on the supposition of the density being precisely as the compression; for by applying the reduction of art. 17, we have, $z=q\left(1+m(\frac{\mathrm{H}+h}{2}-r)\right)\log.\frac{b}{\beta}$. But when n, though very small, does not vanish altogether, by the same reduction,

$$z=q^{1+n}\left(1+m(\frac{\mathrm{H}+h}{2}-r)\right)\log.\frac{b}{\beta}\left(1-\frac{n}{2}\log.b\beta\right).$$

If, therefore, we suppose q^{1+n} to be equal to p, or to 4343 fathoms, which must be nearly true; and, if we call A the height, or the value of z, computed from the formula $z=p\left(1+m(\frac{\mathrm{H}+h}{2}-r)\right)\log.\frac{b}{\beta}$, the correction to be applied on account of n, will be $-\frac{n}{2}\mathrm{A}\log b\beta$.

30. It is not, however, now a matter of indifference in what measure the lengths of the columns of mercury in the barometers are expressed, as it was, when only the ratios of these columns entered into the computation. They must be expressed in terms of the same measure, wherein the height of the mountain is required, and wherein q has been already determined. For, if we take the exact expression for the height, viz.

$$z = \frac{\frac{1}{n} q^{1+n} g(H-h)\left(\frac{1}{\beta^n} - \frac{1}{b^n}\right)}{(1-m)^{H-r} - (1-m)^{h-r}},$$ or that to which it may be reduced, $z = q\left(1 + m\left(\frac{H+h}{2} - r\right)\right)\left(\frac{q^n}{n\beta^n} - \frac{q^n}{nb^n}\right)$, it is evident, that $\frac{q^n}{n\beta^n} - \frac{q^n}{nb^n}$ can have no definite signification, unless b, β, and q be all expressed in terms of the same measure. As the conveniency of computation requires that p or q^{1+n} should be expressed in fathoms, so b and β must also be expressed in parts of a fathom. The same is true of the logarithmic expression, $\frac{n}{2}\log. b\beta$, to which the preceding one is reduced. Thus, if $b=30$ inches, and $\beta=20$ inches, we must make $b=\frac{5}{12}$, and $\beta=\frac{5}{18}$, so that $b\beta = \frac{5\times 5}{12\times 6\times 3}$, half the hyperbolic logarithm of which, or that of $\frac{5}{6\sqrt{6}}$, is $=-1.0782$, and this multiplied

into $-n$, supposing $n=.0015$, gives $+.0016$ to be multiplied into A, or the height as already approximated. The correction here is, therefore, about $\frac{1}{637}$ of A. In other cases, it will exceed this proportion as $b\beta$ diminishes, but (because $b\beta$ will rarely be greater than $\frac{25}{144}$), its minimum will be about $\frac{1}{770}$. In the measurement of great heights, therefore, this equation may deserve to be considered.

31. We come now to find the correction which must be made on the ordinary rule, on account of the diminution of gravity as we ascend from the surface of the earth. By § 27, we have,

$$\frac{q^{1+n}\left(1+\frac{z}{s}\right)^n}{\beta}-\frac{q^{1+n}}{b^n}=\frac{nz}{g(\mathrm{H}-h)}\Big((1-m)^{\mathrm{H}-r}-$$

$$(1-m)^{h-r}-\frac{2z}{gs(\mathrm{H}-h)}(1-m)^{\mathrm{H}-r}+\frac{2z}{s}(1-m)^{h-r}+$$

$$\frac{2z}{gs(\mathrm{H}-h)}(1-m)^{h-r}\Big);$$ and since we know already,

that $z=\dfrac{\frac{1}{n}q^{1+n}g(\mathrm{H}-h)\left(\frac{1}{\beta^n}-\frac{1}{b^n}\right)}{(1-m)^{\mathrm{H}-r}-(1-m)^{h-r}}$ nearly, if we substitute this value of z, or rather that which was before derived from it, viz. $z=q\left(1+m\left(\frac{\mathrm{H}+h}{2}-r\right)\right)$ $\left(\frac{q^n}{n\beta^n}-\frac{q^n}{nb^n}\right)$, in all the terms of this equation, into which s enters as a divisor, we shall have a new and more accurate value of z, and, by a like process, might from thence obtain one still more accurate, if it were necessary.

Now, if this be done, and if the correction depending on n be supposed sufficiently determined by the computations of the two preceding articles, so that it may now be neglected altogether; and if m also be so small, that all the powers of it, higher than the first, may be neglected, we obtain,

$$z=p\left(1+m\left(\frac{H+h}{2}-r\right)\right)\log.\frac{b}{\beta}+\frac{2p^2}{s}\left(1+m\left(\frac{H+h}{2}-r\right)\right)^2 \log.\frac{b}{\beta}+\frac{p^2}{s}\left(1+m\left(\frac{H+h}{2}-r\right)\right)^2\left(\log.\frac{b}{\beta}\right)^2.$$

32. The first term of the preceding equation is the height corrected by M. Deluc's method; the second term, viz. $\frac{2p^2}{s}\left(1+m\left(\frac{H+h}{2}-r\right)\right)^2\log.\frac{b}{\beta}$, is the correction for the diminution of the weight of the quicksilver in the uppermost barometer; and the third term, or $\frac{p^2}{s}\left(1+m\left(\frac{H+h}{2}-r\right)\right)^2\left(\log.\frac{b}{\beta}\right)^2$, is the correction for the gradual diminution of the weight of the air in the different strata between the lower and the upper station. The last of these two corrections, which, in all ordinary cases, is also the least, is the only one of them to which, it would seem, that any attention has hitherto been paid. The other, or the effect of the diminution of the gravity of the quicksilver, was included in this in-

vestigation, when, at § 25, we substituted for y its value, $\frac{\left(\frac{s^{\nu}}{(s+x)^{\nu}}\beta\right)^{1+n}}{q^{1+n}(1+m)^{\tau+\lambda x}}$. It is found by making as s to $p\left(1+m\left(\frac{H+h}{2}-r\right)\right)$, so twice the height, computed by the ordinary method, to a fourth proportional, which is to be added to that height.

The correction for the diminished gravity of the air is a third proportional to the semi-diameter of the earth, and the height, as computed by the ordinary rule. For different mountains, therefore, this correction is in the duplicate ratio of their heights.

These corrections are both additive, and for such a mountain as Coraçon may be equal, the first to 42, and the second to 12 feet.

33. In the measurement of depths below the surface of the earth, β is greater than b, and $=-1$, so that the compressing force, at any depth x, below the surface, is $=\left(\int\frac{s-x}{s}y\dot{x}\right)^{1+n}$, where the fluent is affirmative, not negative, as in all the preceding instances, because the air which, by its weight, compresses the stratum at the depth x, is on the same side of that stratum with x, whereas it was before on the opposite side.

Making, therefore, $y=\frac{+\left(\int\frac{s-x}{s}y\dot{x}\right)^{1+n}}{p(1-m)^{\tau+\lambda x}}$, we have, by proceeding as above,

$$z=p\left(1+m\left(\frac{H+h}{2}-r\right)\right)\log.\frac{\beta}{b}-\frac{p^2}{s}\left(1+m\left(\frac{H+h}{2}-r\right)\right)$$
$$\log.\frac{\beta}{b}+\frac{p^2}{2s}\left(1+m\left(\frac{H+h}{2}-r\right)\right)^2\left(\log.\frac{\beta}{b}\right)^2.$$

In this formula, the second term, viz.

$-\frac{p^2}{s}\left(1+m\left(\frac{H+h}{2}-r\right)\right)\log.\frac{\beta}{b}$ is just half the corresponding term in the preceding formula, (§ 31,) with a contrary sign, so that the correction for the diminution of the gravity of the quicksilver takes away from a depth, as it adds to an elevation. The correction $\frac{p^2}{2s}\left(1+m\frac{H+h}{2}-r\right)\Big)^2\left(\log.\frac{\beta}{b}\right)^2$ retains the same sign in both cases, but in this is only half of what it was in the former. That these last corrections should be each half of the corresponding one in the preceding case, might have been concluded from this, that, by any small ascent above the surface of the earth, the force of gravity is twice as much diminished as by an equal descent below it. The reason of the change of the signs in the second term is also sufficiently obvious.

34. Though these corrections suppose that z is small in respect of s, yet they would afford a sufficient approximation to the truth, were we to reason concerning much greater depths under the surface of the earth than any to which man can penetrate. For example, on a supposition that the atmosphere was continued downwards within the earth, its den-

sity being always as its compression, and its temperature every where the same, (and, for the greater ease of computation equal to r), let it be required to find, at what depth its density would become equal to that of mercury. To resolve this problem, it must be remembered, that the density of mercury, throughout all this computation, has been supposed $=1$, and p equal to the height of a column of mercury, which, gravitating every where with the same force as at the surface, would, by its pressure, give to air the density 1. If a barometer, therefore, were carried down to the depth at which air was as dense as mercury, the mercury in it would rise to the height p, or to 4343 fathoms nearly, supposing, at the same time, that its own gravity were not diminished. Now, on this supposition, (by § 33,) any depression below the surface, as, $z=$ $p \log.\frac{\beta}{b}+\frac{p^2}{2s}\left(\log.\frac{\beta}{b}\right)^2$, the temperature being supposed $=r$, and the term $-\frac{p^2}{s}\log.\frac{\beta}{b}$ being left out, as relating only to the diminution of the weight of the quicksilver in the lower barometer. If, then, b, or the column of mercury in the barometer at the surface, be 30 inches, or $\frac{5}{12}$ of a fathom, and $\beta=4343$, we find $p\log.\frac{\beta}{b}=$ 10000 × tabular log.10423= 40180 fathoms =45.6 miles nearly. The second term, $\frac{p^2}{2s}\left(\log.\frac{\beta}{b}\right)^2$, (or the square of the former divided by the diameter of the earth), $=+.25$ of a mile, so that $z=45.85$ miles nearly. The approximation might be carried to much greater exactness if it were necessary; but this is sufficient to show, that, at a less depth under the surface than 46 miles, the density of air would become equal to that of quicksilver; and if this conclusion appear in any degree paradoxical, it need only be considered, that, abstracting from any diminution of the power of gravitation, the density of air would be nearly doubled by every $3\frac{1}{2}$ miles of descent below the surface of the earth.

35. If, again, we would form any conclusion concerning the limit to which our atmosphere may extend upwards, we must resume the formula,

$$y=\frac{\left(-\int\frac{s}{(s+x)^2}y\dot{x}\right)^{1+n}}{q^{1+n}(1-m)^{\tau+\lambda x}};$$

and, if we would abstract from the effect of the cold in the higher regions to reduce the atmosphere within narrower limits than those to which it would otherwise extend, we may suppose the temperature $r+f$ to be uniformly diffused through it, and so for $(1-m)^{\tau+\lambda x}$ we may substitute $1+fm$. Putting also $a=q(1+fm)^{\frac{1}{1+n}}$, and making $s+x$, or the distance from the centre, $=v$, $ay^{\frac{1}{1+n}}=-\int s^2v^{-2}y\dot{v}$;

wherefore, taking the fluxions, dividing by y, and integrating,

$$-\frac{a}{n}y^{-\frac{n}{1+n}}+C=(\nu-1)s^{\nu}v^{1-\nu}$$

To define C, suppose that $y=D$ when $x=0$, or when $v=s$; then, $C=\frac{a}{n}D^{-\frac{n}{1+n}}+(\nu-1)s$; and so,

$$\frac{a}{n}\left(D^{-\frac{n}{1+n}}-y^{-\frac{n}{1+n}}\right)=(\nu-1)(s^{\nu}v^{1-\nu}-s);$$

and making $\nu=2$, $\frac{a}{n}\left(D^{-\frac{n}{1+n}}-y^{-\frac{n}{1+n}}\right)=s\left(\frac{s}{v}-1\right)$.

36. Now, if n be affirmative, as has been supposed, this formula, because of the negative exponent of y, gives s infinite when $y=0$. The atmosphere, therefore, on this supposition, admits of no limit. But, if we suppose n to be negative, that is, if we suppose the density to be as the power $1-n$ of the compression, instead of $1+n$, the formula of the last article becomes

$$\frac{a}{n}\left(D^{\frac{n}{1-n}}-y^{\frac{n}{1-n}}\right)=s\left(1-\frac{s}{v}\right).$$

And if we now suppose the atmosphere to terminate, or y to become $=0$, then

$\frac{aD^{\frac{n}{1-n}}}{n}=s\left(1-\frac{s}{v}\right)$, and the entire height of the atmosphere, or $v=\frac{s^2}{s-\frac{a}{n}D^{\frac{n}{1-n}}}$.

This value of v may either be finite, infinite, or negative, according to the different magnitudes assigned to n and D. If these be such that s is equal to $\frac{a}{n}\mathrm{D}^{\frac{n}{1-n}}$, it is obvious that v is infinite; but if s be greater than $\frac{a}{n}\mathrm{D}^{\frac{n}{1-n}}$, v must be finite and affirmative. If s be less than $\frac{a}{n}\mathrm{D}^{\frac{n}{1-n}}$, then v is negative; by which we are to understand, that the height of the atmosphere is, as it were, more than infinite, or that its density is finite, even at an infinite distance. It must be remarked, too, that, when n is very small, as it must be in the case of the earth's atmosphere, $\mathrm{D}^{\frac{n}{1-n}}$ being nearly $=1$, we have $v=\frac{s^2}{s-\frac{a}{n}}$. As $a=4343$ fathoms, (on the supposition that the temperature of the atmosphere is 32°,) and as $s=3491840$, it follows, from this formula, that, according as n is greater than .00125, equal to it, or less, the density of the atmosphere will vanish at a finite, an infinite, or not even at an infinite distance.

37. But to return to what is the more immediate object of this paper, it will now be proper to bring into one view the different corrections that

have been investigated. We must, therefore, recollect, that the coefficient p is the length of a column of mercury, which, pressing on air of the temperature r, would give to it the density of mercury, (which is denoted by unity,) supposing, at the same time, that the density of air is as the force compressing it. Hence p is likewise the height of a homogeneous column of air, of any density whatever, which, by its pressure, would make air of the same density with itself; or it is the height to which the atmosphere would extend above the surface of the earth, if it were reduced to the same density throughout, which it has at the surface of the earth, when it is of the temperature r. It has been found by experiment, that, when $r=32°$, p is nearly equal to 4342.9448 fathoms, which number is the modulus of the tabular logarithms multiplied by 10000. This determination, however, is only to be considered as approaching to the truth, if we are to have regard to the following corrections. Instead of p, in some of these investigations, we have used q to denote the height of a column of mercury, which, supposing the condensation of air to be as the power $1+n$ of the compressing force, would, by its pressure, give to air the density of mercury, or the density 1; q^{1+n} cannot differ much from p, but its precise length is to be determined only by experiment. In what follows, p is put for the numeral coefficient, whatever it may be, by which the formula must be multiplied to give the height in fathoms, or in any known measure.

The expansion of air for one degree of heat, the temperature being 32°, and the height of the barometer 29.5 inches, is $=m=.00245$ nearly. μ is the exponent of a power such that 29.5 being denoted by γ, $\frac{\beta^{\mu}}{\gamma^{\mu}}\times m=$ the expansion for one degree of heat, when the mercury in the barometer stands at β. The value of μ is not certainly known; it is probably between 1 and $\frac{1}{3}$. n is a number such, that the density of air is as the power $1+n$ of the compressing force; it is supposed $=.0015$.

The heights of the mercury in the barometers, at the lower and upper stations, are b and β; H and h are the temperatures, marked by Fahrenheit's thermometer at those stations respectively, and $\frac{\mathrm{H}+h}{2}-r$ is put $=f$.

39. Then, the first approximation to the height, without any correction, is, $z=p\log.\frac{b}{\beta}$.

1*mo*. The first correction, M. Deluc's, (§ 17,)

$$+m\left(\frac{\mathrm{H}+h}{2}-r\right)p\log.\frac{b}{\beta}.$$

2*do*. The correction for the decrease of heat in the superior strata of the atmosphere, and for the first inequality of expansion, (§ 19,)=

$$+m^2\left(\frac{r}{2}-\frac{H+h}{4}+\frac{r(r-H-h)}{2}+\frac{H^2+4Hh+h^2}{12}\right)p\log.\frac{b}{\beta}.$$

3tio. The correction for the second inequality of expansion, or for its variation by a given change of temperature, according to the pressure, (§ 22,)=

$+\frac{\mu m}{2}\left(\frac{H+h}{2}-r\right)p\log.\frac{b}{\beta}\times\log.\frac{b\beta}{\gamma^2}$; or, if E be put for M. Deluc's, or the first equation, this last =

$+\frac{\mu E}{2}\log.\frac{b\beta}{\gamma^2}$. But as μ does not appear to be very small, it will be more accurate to compute,

$\frac{pfm(b^{\mu}-\beta^{\mu})}{\mu\gamma^{\mu}}$, which includes in it both the first and third corrections, (§ 21.)

4to. The correction on account of the departure of the law of the elasticity of air, from that of the direct ratio of the density, (§ 29,)

$=-\frac{n}{2}p\left(1+m\left(\frac{H+h}{2}-r\right)\right)\log.\frac{b}{\beta}\times\log.b\beta$. In this equation, b and β must be expressed in the same measure with p, that is, in fathoms.

5to. For the diminution of the weight of the quicksilver in the upper barometer, there is an equation to be applied $=+\frac{2p^2}{s}\left(1+m\left(\frac{H+h}{2}-r\right)\right)^2\log\frac{b}{\beta}$.

6to. On account of the diminished gravity of the air in ascending from the surface of the earth, there is a sixth correction $=+\frac{p^2}{s}\left(1+m\left(\frac{H+h}{2}-r\right)\right)^2$ $\left(\log.\frac{b}{\beta}\right)^2$.

When a depth below the surface is to be measured, the fifth equation becomes negative, and loses the multiplier 2; the sixth remains affirmative, but is divided by 2.

40. These equations, even exclusive of the first, may, in the measurement of great heights, amount to a considerable proportion of the whole. In the instance of Coraçon, 15833 feet above the level of the sea, the greatest height to which the barometer has ever been carried, the first equation exceeds 1100 feet, and the third appears not to be less than — 300. The remaining corrections are, indeed, less considerable; but, being all affirmative, they must not be entirely neglected. And, on the whole, it is certain, that, though the first equation alone will give the height sufficiently exact, while it does not exceed five or six thousand feet, yet, at greater elevations, the corrections that have now been enumerated must all be taken into account. To facilitate the computation by means of them, they ought to be reduced into tables adjusted to their proper arguments, after the values of p, m and r are accurately determined, by comparing the formula that has been given here with observations. But this would lead into disquisitions far exceeding

the bounds of the present inquiry, the object of which is, to ascertain the form, rather than the absolute quantity of these corrections.

41. It is evident, that, in the preceding investigation, as well as in all the other methods of measuring heights by the barometer, it is supposed, either that the one of the barometers is vertical to the other, or that a perfect equilibrium prevails through that part of the atmosphere intercepted between them. The determination of the constant quantity in the foregoing integrations, by supposing that $b=\beta$ when $x=0$, or that the mercury in the two barometers stands at the same height in them, when they are at the same distance from the surface of the earth, obviously involves in it either the one or the other of these conditions. But the last of them, the equilibrium of the atmosphere, never takes place; and, therefore, it is necessary, in order that barometrical measurements be perfectly accurate, that the one barometer be immediately above the other, or, at least, that the horizontal distance between them be very small. If this be not the case, the unequal distribution of the heat through the different parts of the same stratum of air will render it impossible to deduce the difference of the heights of the barometers from a comparison of the columns of mercury contained in them.

For instance, let there be three barometers; the *first* at the surface of the earth, the *second* raised up into the air perpendicularly above the *first*, and the *third* removed into a colder climate, but raised up also into the air, so as to have in it a column of mercury of the same length with that in the *second*. These two last, when compared together by M. Deluc's, or by the preceding rules, will appear to be at the same height above the surface, or above the first barometer. But, if each of them be compared with the *first*, the *second* will appear more elevated above it than the *third*, because of the greater cold supposed to prevail in the region where this last barometer is placed. Here, therefore, are two different determinations of the height of the third station above the first, neither of which has any claim to be preferred to the other. It is evident, therefore, that, in barometrical measurements, there is always a degree of uncertainty introduced by the horizontal distance between the two stations, and that, beside those accidental errors, which are of the less consequence, that, in a number of observations, they may nearly compensate for one another.

It must be confessed, too, that we have not at present the means of removing this uncertainty, nor even of ascertaining its limits with tolerable exactness. These depend on a problem which is no longer to be resolved by the principles of statics, but requires the *motions* of an elastic fluid, under various degrees of compression and rarefaction, to be determined. The solution, therefore, is extremely difficult; and no result, sufficiently simple to be of use in these computations, is ever likely to be obtained from it.

It would, however, be of consequence to determine, by observation, the mean height of the barometer at the level of the sea in the different regions of the earth. That mean height is not every where the same. Under the line, it appears, from the observations of M. Bouguer, to be 29.852 inches, reducing the mercury to the temperature of 55°; and in Britain, it is 30.04, reducing the mercury to the same temperature. The mean temperature of the air, as well as its mean weight in different climates, will also require to be determined before the art of levelling extensive tracts by the barometer can be brought to perfection.

42. There is another cause of error which, had the effects of it been sufficiently known, ought, no doubt, to have entered into this investigation. Moisture, when chemically united to air, or dissolved in it, so as to compose a part of the same homogeneous and invisible fluid, appears to have a powerful effect to increase the elasticity of the air, and its expansion for every additional degree of heat which it receives. In experiments with the manometer, * it has been observed, that, till the

* See General Roy's Experiments, § 2, Phil. Trans. Vol. LXVII. Part II.

moisture was dissolved in the air, it had no sensible effect on its elasticity; but that as soon as it began to dissolve, the expansion, for one degree of heat, was increased, and continued to be so, for every successive addition of heat, from thence to the boiling point, where it became nine times that of dry air. From this, too, it probably proceeded, that, at Spitzbergen, within ten degrees of the pole, a place where the circle of perpetual congelation in the atmosphere, approaches near to the surface of the earth, and where the air may naturally be supposed to be very dry, the usual rule for the measurement of heights was found to err greatly in excess, and it appeared, that the density of the air was greater than could have been inferred from its compression and its temperature.

43. Though the judicious and accurate experiments of General Roy have ascertained this effect of humidity, and have even gone far to determine the law of its operation, yet, for want of a measure of the quantity of it, contained, at any given time, in the air, it is impossible to make any application of this knowledge to the object under our consideration. While I was reflecting on this difficulty, it occurred, that the barometer itself might become a measure of the humidity of the air, and that the error committed in the measuring of a known height, if all other circumstances were taken in,

would determine the quantity of that humidity. For, if we suppose, that the formula

$z=p\left(1+m(\frac{H+h}{2}-r)\right)\log.\frac{b}{\beta}$ gives the true height between the stations at which two barometers have been observed, when the moisture dissolved in the air is of its medium quantity, (which we may call unity,) then, if that moisture be either increased or diminished, the expression $p\left(1+m(\frac{H+h}{2}-r)\right)\log.\frac{b}{\beta}$ will no longer be equal to the true height, but must be multiplied into $1\pm\pi$ in order that it may be equal to z. Now, this fraction $\pm\pi$ represents the excess or defect of the moisture dissolved in the air above or below its mean quantity; or, more exactly, it is proportional to the increase or diminution of the elasticity of the air arising from that cause. When $p\left(1+m(\frac{H+h}{2}-r)\right)\log.\frac{b}{\beta}$ is less than the true height, the fraction π must be affirmative, and indicates an increase of elasticity, and, consequently, of moisture in the air. The contrary happens when $p\left(1+m(\frac{H+h}{2}-r)\right)\log.\frac{b}{\beta}$ is greater than the true height. To determine π, since $z=(1+\pi)p\left(1+m(\frac{H+h}{2}-r)\right)\log.\frac{b}{\beta}$, $1+\pi=\frac{z}{p\left(1+m(\frac{H+h}{2}-r)\right)\log.\frac{b}{\beta}}$. Or if the error, that is

$$z-p\left(1+m\left(\frac{H+h}{2}-r\right)\right)\log.\frac{b}{\beta}=e, \text{ then}$$

$$\pi=\frac{+e}{p\left(1+m\left(\frac{H+h}{2}-r\right)\right)\log.\frac{b}{\beta}}, \text{ or } \pi=\frac{+e}{z-e}.$$

44. To apply the barometer, therefore, for the purposes of hygrometry, let there be two barometers fixed, the one at the top, and the other at the bottom of a high tower, or hill of moderate elevation, and let them be observed at the same instant, together with their corresponding thermometers. If the difference of their heights, computed from thence, be equal precisely to the true difference, then is the moisture dissolved in the air no way different from its mean quantity; but if the difference of the heights so computed be greater or less than the truth, then π, as above determined, will give the quantity by which the actual moisture in the air is less or greater than the mean quantity. The height at which the one barometer should be placed above the other, ought not to be so small that the unavoidable errors of observation, (which may amount to five feet,) may be considerable in respect of the whole; nor so great as to introduce error from other causes. It ought not, therefore, to be less than 100, nor much greater than 500 feet.

45. In this manner, we shall have a measure, not indeed of the absolute quantity of humidity

dissolved in the air at a given time, but of the differences of the humidity dissolved in it at different times. Our hygrometer, therefore, will afford a scale for the measuring of moisture, not unlike that which the thermometer affords for the measuring of heat; and both deduced from the changes produced on the bulk, or the specific gravity of certain bodies. The beginning, or zero, of this scale may also be fixed by a certain and invariable rule, if we assume m, in the preceding formula, (or the expansion of air for one degree of heat,) of a given magnitude, as, for instance, .00245, and conceive the scale to begin when $\pi = 0$, or when the formula, thus adjusted, gives the true height.

The hygrometer with which we will be thus furnished, seems well adapted to the purposes of astronomy. For it measures the humidity chemically united with the air, and not merely the disposition of the air to deposit that humidity, which, though much connected with the changes of the weather, has little to do with the astronomical refraction. It is true, that the fractions π may not be directly proportional to the differences of the humidity of the air, nor to the changes of refracting power, which those differences of humidity may produce; but they are probably connected with these last, by some fixed and invariable law, which future experiments may be able to ascertain. Nor can this application of the barometer fail of leading to some useful conclusion ; for if, on trial, it shall be found, that the operation of humidity in changing the specific gravity of the air, is overruled or concealed by the action of more powerful causes, the discovery, even of this fact, will give a value to the observations.

FINIS.

REMARKS

ON THE ASTRONOMY OF

THE BRAHMINS.

ASTRONOMY

OF THE BRAHMINS.*

1. SINCE the time when astronomy emerged from the obscurity of ancient fable, nothing is better known than its progress through the different nations of the earth. With the era of Nabonassar, regular observations began to be made in Chaldea; the earliest which have merited the attention of succeeding ages. The curiosity of the Greeks was, soon after, directed to the same object; and that ingenious people was the first that endeavoured to explain, or connect by theory, the various phenomena of the heavens. This work was supposed to be so fully accomplished in the Syntaxis of Ptolemy, that his system, without opposition or improvement, continued, for more than five hundred years, to direct the astronomers of Egypt, Italy, and Greece. After the sciences were banished from Alexandria, his writings made their

* From the Transactions of the Royal Society of Edinburgh, Vol. II. (1790.)—ED.

way into the east, where, under the Caliphs of Bagdat, astronomy was cultivated with diligence and success. The Persian princes followed the example of those of Bagdat, borrowing besides, from Trebizond, whatever mathematical knowledge was still preserved among the ruins of the Grecian empire. The conquests of Gengis, and afterwards of Timour, though they retarded, did not stop the progress of astronomy in the east. The grandsons of these two conquerors were equally renowned for their love of science; Hulagu restored astronomy in Persia, and Ulugh Beigh, by an effort still more singular, established it in Tartary. In the mean time, having passed with the Arabs into Spain, it likewise found, in Alphonso of Castile, both a disciple and a patron. It was carried, soon after, into the north of Europe, where, after exercising the genius of Copernicus, of Kepler, and of Newton, it has become the most perfect of all the sciences.

2. In the progress which astronomy has thus made, through almost all the nations, from the Indus to the Atlantic, there is scarce a step which cannot be accurately traced; and it is never difficult to determine what each age, or nation, received from another, or what it added to the general stock of astronomical knowledge. The various systems that have prevailed in all these countries, are visibly connected with one another; they are all derived from one original, and would incline us to believe

that the manner in which men begin to observe the heavens, and to reason about them, is an experiment on the human race, which has been made but once.

It is, therefore, matter of extreme curiosity to find, beyond the Indus, a system of astronomical knowledge that appears to make no part of the great body of science, which has traversed, and enlightened the other countries of the earth; a system that is in the hands of men, who follow its rules without understanding its principles, and who can give no account of its origin, except that it lays claim to an antiquity far beyond the period to which, with us, the history of the heroic ages is supposed to extend.

3. We owe our first knowledge of this astronomy to M. La Loubere, who, returning, in 1687, from an embassy to Siam, brought with him an extract from a Siamese manuscript, which contained tables, and rules, for calculating the places of the sun and moon.* The manner in which these rules were laid down, rendered the principles, on which they were founded, extremely obscure; and it required a commentator as conversant with astronomical calculation as the celebrated Cassini, to explain the meaning of this curious fragment. After that period, two other sets of astronomical tables

* Mém. de l'Acad. des Sciences, Tom. VIII. p. 281, &c.

were sent to Paris by the missionaries in Hindostan; but they remained unnoticed, till the return of M. Legentil from India, where he had been to observe the transit of Venus in 1769. This academician employed himself, during the long stay, which his zeal for science induced him to make in that country, in acquiring a knowledge of the Indian astronomy. The Brahmins thought they saw, in the business of an astronomer, the marks of a *Cast*, that had some affinity to their own, and began to converse with M. Legentil, more familiarly than with other strangers. A learned Brahmin of Tirvalore, having made a visit to the French astronomer, instructed him in the methods, which he used for calculating eclipses of the sun and moon, and communicated to him the tables and rules, that are published in the Memoirs of the Academy of Sciences, for 1772. Since that time, the ingenious and eloquent author of the History of Astronomy, has dedicated an entire volume to the explanation, and comparison of these different tables, where he has deduced, from them, many interesting conclusions.* The subject indeed merited his attention; for the Indian astronomy has all the precision necessary for resolving the great questions, with respect to its own origin and antiquity,

* Traité de l'Astronomie Indienne et Orientale, par M. Bailly. Paris, 1787.

and is by no means among the number of those imperfect fragments of ancient knowledge, which can lead no farther than conjecture, and which an astronomer would gladly resign to the learned researches of the antiquary, or the mythologist.

4. It is from these sources, and chiefly from the elaborate investigations of the last mentioned work, that I have selected the materials of the paper, which I have now the honour to lay before this Society; and it is perhaps necessary that I should make some apology for presenting here, what can have so little claim to originality. The fact is, that notwithstanding the most profound respect, for the learning and abilities of the author of the *Astronomie Indienne*, I entered on the study of that work, not without a portion of the scepticism, which whatever is new and extraordinary in science ought always to excite, and set about verifying the calculations, and examining the reasonings in it, with the most scrupulous attention. The result was, an entire conviction of the accuracy of the one, and of the solidity of the other; and I then fancied, that, in an argument of such variety, I might perhaps do a service to others, by presenting to them, that particular view of it, which had appeared to me the most striking. Such, therefore, is the object of these remarks; they are directed to three different points: The first is to give a short account of the Indian astronomy, so far as it

is known to us, from the four sets of tables above mentioned; the second, to state the principal arguments, that can be deduced from these tables, with respect to their antiquity; and the third, to form some estimate of the geometrical skill, with which this astronomical system is constructed. In the first, I have followed M. Bailly closely; in the second, though I have sometimes taken a different road, I have always come to the same conclusion; having aimed at nothing so much, as to reduce the reasoning into a narrow compass, and to avoid every argument that is not purely astronomical, and independent of all hypothesis; in the third, I have treated of a question which did not fall within the plan of M. Bailly's work, but have only entered on it at present, leaving to some future opportunity, the other discussions to which it leads.

5. The astronomy of India, as you already perceive, is confined to one branch of the science. It gives no theory, nor even any description of the celestial phenomena, but satisfies itself with the calculation of certain changes in the heavens, particularly of the eclipses of the sun and moon, and with the rules and tables by which these calculations must be performed. The Brahmin, seating himself on the ground, and arranging his shells before him, repeats the enigmatical verses that are to guide his calculation, and from his little tablets of palm leaves, takes out the numbers that are to be

employed in it. He obtains his result with wonderful certainty and expedition; but having little knowledge of the principles on which his rules are founded, and no anxiety to be better informed, he is perfectly satisfied, if, as it usually happens, the commencement and duration of the eclipse answer, within a few minutes, to his prediction. Beyond this his astronomical inquiries never extend; and his observations, when he makes any, go no farther than to determine the meridian line, or the length of the day, at the place where he observes.

The objects, therefore, which this astronomy presents to us, are principally three. 1. Tables and rules for calculating the places of the sun and moon: 2. Tables and rules for calculating the places of the planets: 3. Rules by which the phases of eclipses are determined. Though it is chiefly to the first of these that our attention at present is to be directed, the two last will also furnish us with some useful observations.

6. The Brahmins, like all other astronomers, have distinguished from the rest of the heavens, that portion of them, through which the sun, moon, and planets continually circulate. They divide this space, which we call the zodiac, into twenty-seven equal parts, each marked by a group of stars, or a constellation.* This division of the zodiac is ex-

* Mém. sur l'Astronomie des Indiens, par M. Legentil,

tremely natural in the infancy of astronomical observation; because the moon completes her circle among the fixed stars, nearly in twenty-seven days, and so makes an actual division of that circle into twenty-seven equal parts. The moon, too, it must be remembered, was, at that time, the only instrument, if we may say so, by which the positions of the stars on each side of her path could be ascertained; and when her own irregularities were unknown, she was, by the rapidity of her motion eastward, well adapted for this purpose. It is also to the phases of the moon, that we are to ascribe the common division of time into weeks, or portions of seven days, which seems to have prevailed almost over the whole earth.* The days of the week are dedicated by the Brahmins, as by us, to the seven planets, and what is truly singular, they are arranged precisely in the same order.

7. With the constellations, that distinguish the twenty-seven equal spaces, into which their zodiac is divided, the astronomers of India have connected none of those figures of animals, which are among us, of so ancient, and yet so arbitrary an original. M. Legentil has given us their names, and confi-

---

Hist. de l'Acad. des Scienc. 1772, II. p. 207. The phrase which we here translate *constellations,* signifies *the places of the moon in the twelve signs.*

* Mém. Acad. des Scienc. 1772, II. p. 189.

gurations.* They are formed, for the most part, of small groups of stars, such as the Pleiades or the Hyades, those belonging to the same constellation being all connected by straight lines. The first of them, or that which is placed at the beginning of their zodiac, consists of six stars, extending from the head of Aries to the foot of Andromeda, in our zodiac, and occupying a space of about ten degrees in longitude. These constellations are far from including all the stars in the zodiac. M. Legentil remarks, that those stars seem to have been selected, which are best adapted for marking out, by lines drawn between them, the places of the moon in her progress through the heavens.

At the same time that the stars in the zodiac are thus arranged into twenty-seven constellations, the ecliptic is divided, as with us, into twelve signs of thirty degrees each. This division is purely ideal, and is intended merely for the purpose of calculation. The names and emblems by which these signs are expressed are nearly the same as with us;† and as there is nothing in the nature of things to have determined this coincidence, it must, like the arrangement of the days of the week, be the result of some ancient and unknown communication.

---

* Mém. Acad. des. Scienc. 1772, II. p. 209.

† Mém. Acad. des Scienc. 1772, II. p. 200. The zodiac they call *sodi mandalam,* or the circle of stars.

8. That motion by which the fixed stars all appear to move eastward, and continually to increase their distance from the place that the sun occupies at the vernal equinox, is known to the Brahmins, and enters into the composition of all their tables. * They compute this motion to be at the rate of 54″ a-year; so that their *annus magnus*, or the time in which the fixed stars complete an entire revolution, is 24000 years. This motion is too quick by somewhat less than 4″ a-year; an error that will not be thought great, when it is considered, that Ptolemy committed one of 14″, in determining the same quantity.

Another circumstance, which is common to all the tables, and, at the same time, peculiar to the Indian astronomy, is, that they express the longitude of the sun and moon, by their distance from the beginning of the moveable zodiac, and not, as is usual with us, by their distance from the point of the vernal equinox. The longitude is reckoned in signs of 30°, as already mentioned, and each degree is subdivided into 60′, &c. In the division of time, their arithmetic is purely sexagesimal: They divide the day into 60 hours, the hour into 60 minutes, &c.; so that their hour is 24 of our minutes, their minute 24 of our seconds, and so on.

---

* Mém. Acad. des Scienc. 1772, II. p. 194. Astr. Indienne, p. 43, &c.

9. These remarks refer equally to all the tables. We are now to take notice of what is peculiar to each, beginning with those of Siam.

In order to calculate for a given time, the place of any of the celestial bodies, three things are requisite. The first is, the position of the body in some past instant of time, ascertained by observation; and this instant, from which every calculation must set out, is usually called the *epoch* of the tables. The second requisite is, the mean rate of the planet's motion, by which is computed the arch in the heavens, that it must have described, in the interval between the epoch and the instant for which the calculation is made. By the addition of this, to the place at the epoch, we find the mean place of the planet, or the point it would have occupied in the heavens, had its motion been subject to no irregularity. The third is, the correction, on account of such irregularity, which must be added to the mean place, or subtracted from it, as circumstances require, in order to have the true place. The correction thus made is, in the language of astronomy, called an equation; and, when it arises from the eccentricity of a planet's orbit, it is called the equation of the centre.

10. The epoch of the tables of Siam does not go back to any very remote period. Cassini, by an ingenious analysis of their rules, finds that it corresponds to the 21st of March, in the year 638 of

our era, at three in the morning, on the meridian of Siam.* This was the instant at which the astronomical year began, and at which both the sun and moon entered the moveable zodiac. Indeed, it is to be observed, that, in all the tables, the astronomical year begins when the sun enters the moveable zodiac, so that the beginning of this year is continually advancing with respect to the seasons, and makes the complete round of them in 24000 years.

From the epoch above mentioned, the mean place of the sun for any other time is deduced, on the supposition that in 800 years, there are contained 292207 days. † This supposition involves in it the length of the sidereal year, or the time that the sun takes to return to the beginning of the moveable zodiac, and makes it consist of *365 d.* 6 *h.* 12′, 36″. ‡ From this, in order to find the tropical year, or that which regulates the seasons, we must take away 21 , *55*″, as the time which the sun takes to move over the *54*″, that the stars are supposed to have advanced in the year; there will remain 365 *d.* *5 h.* 50′, 41″, which is the length of the tropical year that is involved, not only in the tables of Siam,

* Mém. Acad. Scienc. Tom. VIII. p. 312. Astr. Indienne, p. 11, § 14.

† Astr. Indienne, p. 7, § 6.

‡ Mém. Acad. Scienc. Tom. VIII. p. 328.

but likewise, very nearly, in all the rest.* This determination of the length of the year is but 1, 53″, greater than that of Lacaille, which is a degree of accuracy beyond what is to be found in the more ancient tables of our astronomy.

11. The next thing with which these tables present us, is a correction of the sun's mean place, which corresponds to what we call the equation of his centre, or the inequality arising from the eccentricity of his orbit, in consequence of which, he is alternately retarded and accelerated, his true place being, for one half of the year, left behind the mean, and, for the other, advanced before it. The point where the sun is placed, when his motion is slowest, we call his apogee, because his distance from the earth is then greatest; but the Indian astronomy, which is silent with respect to theory, treats this point as nothing more than what it appears to be, a point, viz. in the heavens, where the sun's motion is the slowest possible, and about 90° distant from that, where his greatest inequality takes place. This greatest inequality is here made to be 2°, 12′,† about 16′ greater than it is deter-

* Astr. Indienne, p. 124. The tables of Tirvalore make the year 6″ less.

† The equation of the sun, or what they call the *chaiaa*, is calculated in the Siamese tables only for every 15° of the *matteiomme*, or mean anomaly. Cassini, *ubi supra*, p. 299.

mined, by the modern astronomy of Europe. This difference is very considerable; but we shall find that it is not to be ascribed wholly to error, and that there was a time when the inequality in question was nearly of the magnitude here assigned to it. In the other points of the sun's path, this inequality is diminished, in proportion to the sine of the mean distance from the apogee, that is, nearly as in our own tables. The apogee is supposed to be 80° advanced beyond the beginning of the zodiac, and to retain always the same position among the fixed stars, or to move forward at the same rate with them. * Though this supposition is not accurate, as the apogee gains upon the stars about 10″ annually, it is much nearer the truth than the system of Ptolemy, where the sun's apogee is supposed absolutely at rest, so as continually to fall back among the fixed stars, by the whole quantity of the precession of the equinoxes. †

12. In these tables, the motions of the moon are deduced, by certain intercalations, from a period of nineteen years, in which she makes nearly 235 re-

---

* Astr. Indienne, p. 9.

† The error, however, with respect to the apogee, is less than it appears to be; for the motion of the Indian zodiac, being nearly 4″ swifter than the stars, is but 6″ slower than the apogee. The velocity of the Indian zodiac is, indeed, neither the same with that of the stars, nor of the sun's apogee, but nearly a mean between them.

volutions; and it is curious to find at Siam, the knowledge of that cycle, of which the invention was thought to do so much honour to the Athenian astronomer Meton, and which makes so great a figure in our modern calendars.* The moon's apogee is supposed to have been in the beginning of the moveable zodiac, 621 days after the epoch of the 21st of March 638, and to make an entire revolution in the heavens in the space of 3232 days.† The first of these suppositions agrees with Mayer's tables to less than a degree, and the second differs from them only by 11 *h.* 14′, 31″; and if it be considered that the apogee is an ideal point in the heavens, which even the eyes of an astronomer cannot directly perceive, to have discovered its true motion so nearly, argues no small correctness of observation.

13. From the place of the apogee, thus found, the inequalities of the moon's motion, which are to reduce her mean to her true place, are next to be determined. Now, at the oppositions and conjunctions, the two greatest of the moon's inequalities, the equation of the centre and the evection, both depend on the distance from the apogee, and

* The Indian period is more exact than that of our golden number by 35′. Astr. Indienne, p. 5. The Indians regulate their festivals by this period. Ibid. Disc. Prelim. p. 8.

† Astr. Indienne, p. 11 and 20.

therefore appear but as one inequality. They also partly destroy one another; so that the moon is retarded or accelerated, only by their difference, which, when greatest, is, according to Mayer's tables, 4°, 57′, 42″. The Siamese rules, which calculate only for oppositions and conjunctions, give, accordingly, but one inequality to the moon, and make it, when greatest, 4°, 56′, not 2′ less than the preceding. This greatest equation is applied, when the moon's mean distance from the apogee is 90°; in other situations, the equation is less, in proportion as the sine of that distance diminishes.*

14. The Siamese MS. breaks off here, and does not inform us how the astronomers of that country proceed, in the remaining parts of their calculation, which they seem to have undertaken, merely for some purpose in astrology. Cassini, to whom we are indebted for the explanation of these tables, observes, that they are not originally constructed for the meridian of Siam, because the rules direct to take away 3′ for the sun, and 40′ for the moon, (being the motion of each for 1*h*. 13′,) from their longitudes calculated as above.† The meridian of the tables is therefore 1*h*. 13′, or 18°, 15′, west of Siam; and it is remarkable, that this brings us very near

---

* Astr. Indienne, p. 13. Cassini, Mém. Acad. Tom. VIII. p. 304.

† Mém. Acad. Scienc. Tom. VIII. p. 302 and 309.

to the meridian of Benares, the ancient seat of Indian learning.* The same agrees nearly with what the Hindoos call their first meridian, which passes through Ceylon and the Banks of Ramanancor. We are, therefore, authorized, or rather, we are necessarily determined to conclude, that the tables of Siam came originally from Hindostan.

15. Another set of astronomical tables, now in the possession of the Academy of Sciences, was sent to the late M. De Lisle from Chrisnabouram, a town in the Carnatic, by Father Du Champ, about the year 1750. Though these tables have an obvious affinity to what have already been described, they form a much more regular and extensive system of astronomical knowledge. They are fifteen in number; and include, beside the mean motions of the sun, moon and planets, the equations to the centre of the sun and moon, and two corrections for each of the planets, the one of which corresponds to its apparent, and the other to its real inequality. They are accompanied also with precepts, and examples, which Father Du Champ received from the Brahmins of Chrisnabouram, and which he has translated into French.†

* Astr. Ind. p. 12. It brings us to a meridian 82° 34′, east of Greenwich. Benares is 83° 11′, east of the same, by Rennell's map.

† These tables are published by Bailly, Astr. Ind. p. 335, &c. See also p. 31, &c.

The epoch of these tables is less ancient than that of the former, and answers to the 10th of March at sunrise, in the year 1491 of our era, when the sun was just entering the moveable zodiac, and was in conjunction with the moon; two circumstances, by which almost all the Indian eras are distinguished. The places, which they assign, at that time, to the sun and moon, agree very well with the calculations made from the tables of Mayer, and Lacaille. In their mean motions, they indeed differ somewhat from them; but as they do so equally for the sun and moon, they produce no error, in determining the relative position of these bodies, nor, of consequence, in calculating the phenomena of eclipses. The sun's apogee is here supposed to have a motion swifter than that of the fixed stars, by about 1″ in nine years, which, though it falls greatly short of the truth, does credit to this astronomy, and is a strong mark of originality. The equation of the sun's centre is somewhat less here than in the tables of Siam; it is 2°, 10′, 30″; the equation of the moon's centre is 5°, 2′, 47″; her path, where it intersects that of the sun, is supposed to make an angle with it of 4°, 30′, and the motions, both of the apogee and node, are determined very near to the truth.

16. Another set of tables, sent from India by Father Patouillet, were received by M. De Lisle,

about the same time with those of Chrisnabouram. They have not the name of any particular place affixed to them; but, as they contain a rule for determining the length of the day, which answers to the latitude of 16° 16′, Bailly thinks it probable that they come from Narsapour.*

The precepts and examples, which accompany these tables, though without any immediate reference to them, are confined to the calculation of the eclipses of the sun and moon; but the tables themselves extend to the motion of the planets, and very much resemble those of Chrisnabouram, except that they are given with less detail, and in a form much more enigmatical.† The epoch of the precepts, which Bailly has evolved with great ingenuity, goes back no farther than the year 1569, at midnight, between the 17th and 18th of March. From this epoch, the places of the sun and moon are computed, as in the tables of Siam, with the ad-

---

* Astr. Ind. p. 49, &c.

† They were explained, or rather decyphered by M. Legentil in the Memoirs of the Academy of Sciences for 1784, p. 482, &c.; for they were not understood by the missionary who sent them to Europe, nor probably by the Brahmins who instructed him. M. Legentil thinks that they have the appearance of being copied from inscriptions on stone. The minutes and seconds are ranged in rows under one another, not in vertical columns, and without any title to point out their meaning, or their connection. These tables are published, Mém. Acad. *ibid.* p. 492, and Astr. Ind. p. 414.

dition of an equation, which is indeed extremely singular. It resembles that correction of the moon's motion, which was discovered by Tycho, and which is called the annual equation, because its quantity depends, not on the place of the moon, but on the place of the sun, in the ecliptic. It is every where proportional to the inequality of the sun's motion, and is nearly a tenth part of it. The tables of Narsapour make their annual equation only $\frac{1}{27}$ of the sun's: but this is not their only mistake; for they direct the equation to be added to the moon's longitude, when it ought to be subtracted from it, and *vice versa*. Now, it is difficult to conceive from whence the last mentioned error has arisen; for though it is not at all extraordinary, that the astronomers, who constructed these tables, should mistake the quantity of a small equation, yet it is impossible, that the same observations, which informed them of its existence, should not have determined, whether it was to be added or subtracted. It would seem, therefore, that something accidental must have occasioned this error; but however that be, an inequality in the lunar motions, that is found in no system with which the astronomers of India can have had any communication, is at least a proof of the originality of their tables.

17. The tables, and methods, of the Brahmins of Tirvalore, are, in many respects, more singular

than any that have yet been described.* The solar year is divided, according to them, into twelve unequal months, each of which is the time that the sun takes to move through one sign, or 30°, of the ecliptic. Thus, *Any*, or June, when the sun is in the third sign, and his motion slowest, consists of 31*d.* 86*h.* 38′, and *Margagy*, or December, when he is in the ninth sign, and his motion quickest, consists only of 29*d.* 20*h.* 53′.† The lengths of these months, expressed in natural days, are contained in a table, which, therefore, involves in it the place of the sun's apogee, and the equation of his centre. The former seems to be 77° from the beginning of the zodiac, and the latter about 2°, 10′, nearly as in the preceding tables. In their calculations, they also employ an astronomical day, which is different from the natural, being the time that the sun takes to move over one degree of the eclip-

* Tirvalore is a small town on the Coromandel coast, about 12 G. miles west of Negapatnam, in lat. 10° 44′, and east long. from Greenwich, 79° 42′, by Rennell's map. From the observations of the Brahmins, M. Legentil makes its lat. to be 10° 42′ 13″. (Mém. Acad. Scienc. II. p. 184.) The meridian of Tirvalore nearly touches the west side of Ceylon, and therefore may be supposed to coincide with the first meridian, as laid down by Father Du Champ. There is no reduction of longitude employed in the methods of Tirvalore.

† These are Indian hours, &c.

tic; and of which days there are just 360 in a year.*

18. These tables go far back into antiquity. Their epoch coincides with the famous era of the Calyougham, that is, with the beginning of the year 3102 before Christ. When the Brahmins of Tirvalore would calculate the place of the sun for a given time, they begin by reducing into days the interval between that time, and the commencement of the Calyougham, multiplying the years by 365*d.* 6*h.* 12′, 30″; and taking away 2*d.* 3*h.* 32′, 30″, the astronomical epoch having begun that much later than the civil.† They next find, by means of certain divisions, when the year current began, or how many days have elapsed since the beginning of it, and then, by the table of the duration of months, they reduce these days into astronomical months, days, &c. which is the same with the signs, degrees, and minutes of the sun's longitude from the beginning of the zodiac. The sun's longitude, therefore, is found.

19. Somewhat in the same manner, but by a rule still more artificial and ingenious, they deduce the place of the moon, at any given time, from her

---

* Mém. Acad. des Scienc. II. p. 187. Astr. Indienne, p. 76, &c.

† The Indian hours are here reduced to European.

place at the beginning of the Calyougham.* This rule is so contrived, as to include at once the motions both of the moon and of her apogee, and depends on this principle, according to the very skilful interpretation of Bailly, that, 1600894 days after the above mentioned epoch, the moon was in her apogee, and 7ˢ, 2°, 0′, 7″, distant from the beginning of the zodiac; that after 12372 days, the moon was again in her apogee, with her longitude increased, 9ˢ, 27°, 48′, 10″; that in 3031 days more, the moon is again in her apogee, with 11ˢ, 7°, 31′, 1″, more of longitude; and, lastly, that, after 248 days, she is again in her apogee, with 27°, 44′, 6″, more of longitude. By means of the three former numbers, they find, how far, at any given time, the moon is advanced in this period of 248 days, and by a table, expressing how long the moon takes to pass through each degree of her orbit, during that period, they find how far she is then advanced in the zodiac.† This rule is strongly marked with all the peculiar characters of the Indian astronomy: It is remarkable for its accuracy, and still more for its ingenuity and refinement; but is not reduced withal, to its ultimate simplicity.

* Mém. Acad. des Scienc. *ibid.* p. 229. Astr. Ind. p. 84.

† M. Legentil has given this table, Mém. Acad. *ibid.* p. 261.

20. The tables of Tirvalore, however, though they differ in form very much from those formerly described, agree with them perfectly in many of their elements. They suppose the same length of the year, the same mean motions, and the same inequalities of the sun and moon, and they are adapted nearly to the same meridian.* But a circum-

---

* The accuracy of the geography of the Hindoos, is in no proportion to that of their astronomy, and, therefore, it is impossible that the identity of the meridians of their tables can be fully established. All that can be said, with certainty, is, that the difference between the meridians of the tables of Tirvalore and Siam is, at most, but inconsiderable, and may be only apparent, arising from an error in computing the difference of longitude between these places. The tables of Tirvalore are for Long. 79°, 42′; those of Siam for 82°, 34′; the difference is 2°, 52′, not more than may be ascribed to an error purely geographical.

As to the tables of Chrisnabouram, they contain a reduction, by which it appears, that the place where they are now used is 45′ of a degree east of the meridian for which they were originally constructed. This makes the latter meridian agree tolerably with that of Cape Comorin, which is in Long. 77°, 32′, 30″, and about half a degree west of Chrisnabouram. But this conclusion is uncertain; because, as Bailly has remarked, the tables sent from Chrisnabouram, and understood by Father Duchamp to belong to that place, are not adapted to the latitude of it, but to one considerably greater, as appears from their rule for ascertaining the length of the day. (Astr. Ind. p. 33.)

The characters, too, by which the Brahmins distinguish

stance in which they seem to differ materially from the rest is, the antiquity of the epoch from which they take their date, the year 3102 before the Christian era. We must, therefore, inquire, whether this epoch is real or fictitious, that is, whether it has been determined by actual observation, or has been calculated from the modern epochs of the other tables. For it may naturally be supposed, that the Brahmins, having made observations in later times, or having borrowed from the astronomical knowledge of other nations, have imagined to themselves a fictitious epoch, coinciding with the celebrated era of the Calyougham, to which, through vanity or superstition, they have referred the places of the heavenly bodies, and have only calculated what they pretend that their ancestors observed.

---

their first meridian, are not perfectly consistent with one another. Sometimes it is described as bisecting Ceylon; and at other times, as touching it on the west side, or even as being as far west as Cape Comorin. Lanka, which is said to be a point in it, is understood, by Fath. Duchamp, to be Ceylon. Bailly thinks that it is the lake Lanka, the source of the Gogra, placed by Rennell, as well as the middle of Ceylon, in Long. 80°, 42′; but, from a Hindoo map, in the Ayeen Akbery, Vol. III. p. 25, Lanka appears to be an island which marks the intersection of the first meridian of the map, nearly that of Cape Comorin, with the equator; and is probably one of the Maldives Islands. See also a note in the Ayeen Akbery, *ibid.* p. 36.

21. In doing this, however. the Brahmins must have furnished us with means, almost infallible, of detecting their imposture. It is only for astronomy, in its most perfect state, to go back to the distance of forty-six centuries, and to ascertain the situation of the heavenly bodies at so remote a period. The modern astronomy of Europe, with all the accuracy that it derives from the telescope and the pendulum, could not venture on so difficult a task, were it not assisted by the theory of gravitation, and had not the integral calculus, after an hundred years of almost continual improvement, been able, at last, to determine the disturbances in our system, which arise from the action of the planets on one another.

Unless the corrections for these disturbances be taken into account, any system of astronomical tables, however accurate at the time of its formation, and however diligently copied from the heavens, will be found less exact for every instant, either before or after that time, and will continually diverge more and more from the truth, both for future and past ages. Indeed, this will happen, not only from the neglect of these corrections, but also from the small errors unavoidably committed, in determining the mean motions, which must accumulate with the time, and produce an effect that becomes every day more sensible, as we retire, on either side, from the instant of observation. For both these reasons, it may be established as a maxim, that, if there be given a system of astronomical tables, founded on observations of an unknown date, that date may be found, by taking the time when the tables represent the celestial motions most exactly.

Here, therefore, we have a criterion, by which we are to judge of the pretensions of the Indian astronomy to so great antiquity. It is true, that, in applying it, we must suppose our modern astronomy, if not perfectly accurate, at least so exact as to represent the celestial motions, without any sensible error, even for a period more remote than the Calyougham; and this, considering the multitude of observations on which our astronomy is founded, the great antiquity of some of those observations, and the extreme accuracy of the rest, together with the assistance derived from the theory of physical causes, may surely be assumed as a very reasonable postulatum. We begin with the examination of the mean motions.

22. The Brahmins place the beginning of their moveable zodiac, at the time of their epoch, 54° before the vernal equinox, or in the longitude of 10^s, 6°, according to our method of reckoning. Now, M. Legentil brought with him a delineation of the Indian zodiac, from which the places of the stars in it may be ascertained with tolerable exact-

ness.* In particular, it appears, that Aldebaran, or the first star of Taurus, is placed in the last degree of the fourth constellation, or 53°, 20′, distant from the beginning of the zodiac. Aldebaran was therefore 40 before the point of the vernal equinox, according to the Indian astronomy, in the year 3102 before Christ. But the same star, by the best modern observations, was, in the year 1750, in longitude, 2^s, 6°, 17′, 47″; and had it gone forward, according to the present rate of the precession of the equinoxes, 50″½ annually, it must have been, at the era of the Calyougham, 1°, 32′, before the equinox. But this result is to be corrected, in consequence of the inequality in the precession, discovered by Lagrange, † by the addition of 1°, 45′, 22″, to the longitude of Aldebaran, which gives the longitude of that star 13′ from the vernal equinox, at the time of the Calyougham, agreeing, within 53′, with the determination of the Indian astronomy. ‡

This agreement is the more remarkable, that the Brahmins, by their own rules for computing the motion of the fixed stars, could not have assigned

---

Mém. Acad. Scienc. 1772, II. p. 214. Astr. Ind. p. 129.

† Mém. Acad. de Berlin, 1782, p. 287. Astr. Ind. p. 144.

‡ Astr. Ind. p. 130.

this place to Aldebaran for the beginning of the Calyougham, had they calculated it from a modern observation. For as they make the motion of the fixed stars too great by more than 3″ annually, if they had calculated backward from 1491, they would have placed the fixed stars less advanced by 4° or 5°, at their ancient epoch, than they have actually done. This argument carries with it a great deal of force; and even were it the only one we had to produce, it would render it, in a high degree, probable, that the Indian zodiac was as old as the Calyougham.

23. Let us next compare the places of the sun and moon, for the beginning of the Calyougham, as deduced from the Indian and the modern astronomy. And, first, of the sun, though, for a reason that will immediately appear, it is not to be considered as leading to any thing conclusive. Bailly, from a comparison of the tables of Tirvalore with those of Chrisnabouram, has determined the epoch of the former to answer to midnight, between the 17th and 18th * of February of the year 3102 before Christ, at which time the sun was just

---

* Astr. Ind. p. 110. The Brahmins, however, actually suppose the epoch to be 6 hours later, or at sunrise, on the same day. Their mistake is discovered, as has been said, by comparing the radical places in the different tables with one another.

entering the moveable zodiac, and was therefore in longitude 10ˢ, 6°. Bailly also thinks it reasonable to suppose, that this was not the mean place of the sun, as the nature of astronomical tables require, but the true place, differing from the mean, by the equation to the sun's centre at that time.* This, it must be confessed, is the mark of greatest unskilfulness, that we meet with in the construction of these tables. Supposing it, however, to be the case, the mean place of the sun, at the time of the epoch, comes out 10ˢ, 3°, 38′, 13″. Now, the mean longitude of the sun, from Lacaille's tables, for the same time, is 10ˢ, 1°, 5′, 57″, supposing the precession of the equinoxes to have been uniformly at the rate it is now, that is, 50″⅓ annually. But Lagrange has demonstrated, that the precession was less in former ages than in the present; and his formula gives 1°, 45′, 22″, to be added, on that account, to the sun's longitude already found, which makes it 10ˢ, 2°, 51′, 19″, not more than 47′ from the radical place in the tables of Tirvalore. This agreement is near enough to afford a strong proof of the reality of the ancient epoch, if it were not for the difficulty that remains about considering the sun's place as the true, rather than the mean; and, for that reason, I am unwilling that any stress should be laid upon this argument. The

* Astr. Ind. p. 83.

place of the moon is not liable to the same objection.

24. The moon's mean place, for the beginning of the Calyougham, (that is, for midnight between the 17th and 18th of February 3102, A. C. at Benares,) calculated from Mayer's tables, on the supposition that her motion has always been at the same rate as at the beginning of the present century, is 10^s, 0°, 51$'$, 16$''$.* But, according to the same astronomer, the moon is subject to a small, but uniform acceleration, such, that her angular motion, in any one age, is 9$''$ greater than in the preceding, which, in an interval of 4801 years, must have amounted to 5°, 45$'$, 44$''$. This must be added to the preceding, to give the real mean place of the moon, at the astronomical epoch of the Calyougham, which is therefore 10^s, 6°, 37$'$. Now, the same, by the tables of Tirvalore, is 10^s, 6°, 0$'$; the difference is less than two-thirds of a degree, which, for so remote a period, and considering the acceleration of the moon's motion, for which no allowance could be made in an Indian calculation, is a degree of accuracy that nothing but actual observation could have produced.

---

* Astr. Ind. p. 142, &c. The first meridian is supposed to pass through Benares; but even if it be supposed 3° farther west, the difference, which is here 37$'$, will be only increased to 42$'$.

25. To confirm this conclusion, Bailly computes the place of the moon for the same epoch, by all the tables to which the Indian astronomers can be supposed to have ever had access.* He begins with the tables of Ptolemy; and if, by help of them, we go back from the era of Nabonassar, to the epoch of the Calyougham, taking into account the comparative length of the Egyptian and Indian years, together with the difference of meridians between Alexandria and Tirvalore, we shall find the longitude of the sun 10°, 21′, 15″ greater, and that of the moon 11°, 52′, 7″ greater than has just been found from the Indian tables.† At the same time that this shows, how difficult it is to go back, even for a less period than that of 3000 years, in an astronomical computation, it affords a proof, altogether demonstrative, that the Indian astronomy is not derived from that of Ptolemy.

The tables of Ulugh Beig are more accurate than those of the Egyptian astronomer. They were constructed in a country not far from India, and but a few years earlier than 1491, the epoch of the tables of Chrisnabouram. Their date is July 4, at noon, 1437, at Samarcand; and yet they do not agree with the Indian tables, even at the above mentioned epoch of 1491.‡ But, for the year

* Astr. Ind. p. 114. † Ibid. p. 115.
‡ Ibid. p. 117.

3102 before Christ, their difference from them, in the place of the sun, is 1°, 30′, and in that of the moon 6°; which, though much less than the former differences, are sufficient to show, that the tables of India are not borrowed from those of Tartary.

The Arabians employed in their tables the mean motions of Ptolemy; the Persians did the same, both in the more ancient tables of Chrysococca, and the later ones of Nassireddin. * It is, therefore, certain, that the astronomy of the Brahmins is neither derived from that of the Greeks, the Arabians, the Persians, or the Tartars. This appeared so clear to Cassini, though he had only examined the tables of Siam, and knew nothing of many of the great points which distinguish the Indian astronomy from that of all other nations, that he gives it as his opinion, that these tables are neither derived from the Persian astronomy of Chrysococca, nor from the Greek astronomy of Ptolemy; the places they give at their epoch to the apogee of the sun, and of the moon, and their equation for the sun's centre, being very different from both. †

26. But to return to what respects the moon's acceleration; it is plain, that tables, as ancient as

* Astr. Ind. p. 118.

† Mém. Acad. Scienc. Tom. VIII. p. 286.

those of Tirvalore pretend to be, ought to make the mean motion of that planet much slower than it is at present. They do accordingly suppose, in the rule for computing the place of the moon, already described, that her motion for 4383 years, 94 days, reckoned in the moveable zodiac from the epoch of the Calyougham, is 7ˢ, 2°, 0′, 7″, or 9ˢ, 7°, 45′, 1″, when referred to the fixed point of the vernal equinox. Now, the mean motion for the same interval, taken from the tables of Mayer, is greater than this, by 2°, 42′, 4″,* which, though conformable, in general, to the notion of the moon's motion having been accelerated, falls, it must be confessed, greatly short of the quantity which Mayer has assigned to that acceleration. This, however, is not true of all the tables; for the moon's motion in 4383 years, 94 days, taken from those of Chrisnabouram, is 3°, 2′, 10″ less than in the tables of Tirvalore;† from which it is reasonable to conclude, with Bailly, that the former are, in reality, more ancient than the latter, though they do not profess to be so: and hence, also, the tables of Chrisnabouram make the moon's motion less than Mayer's for the above mentioned interval, by 5°, 44′, 14″, which therefore is, according to them, the quantity of the acceleration.

---

* Astr. Ind. p. 145.

† Ibid. p. 126.

27. Now, it is worthy of remark, that if the same be computed on Mayer's principles, that is, if we calculate how much the angular motion of the moon for 4383 years, 94 days, dated from the beginning of the Calyougham, must have been less than if her velocity had been all that time uniform, and the same as in the present century, we shall find it to be 5°, 43′, 7″, an arch which is only 1′, 7″, less than the former. The tables of Chrisnabouram, therefore, agree with those of Mayer, when corrected by the acceleration within 1′, 7″, and that for a period of more than four thousand years. From this remarkable coincidence, we may conclude, with the highest probability, that at least one set of the observations, on which those tables are founded, is not less ancient than the Calyougham; and though the possibility of their being some ages later than that epoch, is not absolutely excluded, yet it may, by strict mathematical reasoning, be inferred, that they cannot have been later than 2000 years before the Christian era.*

* The reasoning here referred to is the following: As the mean motions, in all astronomical tables, are determined by the comparison of observations made at a great distance of time from one another; if x be the number of centuries between the beginning of the present, and the date of the more ancient observations, from which the moon's mean motion in the tables of Chrisnabouram is deduced; and if y denote the same for the more modern observations: then the quantity

28. This last is one of the few coincidences between the astronomy of India and of Europe, which their ingenious historian has left for others to observe. Indeed, since he wrote, every argument, founded

---

by which the moon's motion during the interval $x-y$, falls short of Mayer's, for the same interval, is $(x^2-y^2)9''$.

If, therefore, m be the motion of the moon for a century in the last mentioned tables, $m(x-y)-9''(x^2-y^2)$ will be the mean motion for the interval $x-y$ in the tables of Chrisnabouram. If, then, a be any other interval, as that of 43.83 centuries, the mean motion assigned to it, in these last tables, by the rule of proportion, will be $\frac{ma(x-y)-9''a(x^2-y^2)}{x-y}=$ $ma-9''a(x+y)$. Let this motion, actually taken from the tables, be $=na$, then $ma-na=9''a(x+y)$, or $x+y=\frac{m-n}{9''}=$ 52.19 in the present case. It is certain, therefore, that whatever supposition be made with respect to the interval between x and y, their sum must always be the same, and must amount to 5219 years. But that, that interval may be long enough to give the mean motions with exactness, it can scarcely be supposed less than 2000 years; and, in that case, $x=3609$ years, which, therefore, is its least value. But if 3609 be reckoned back from 1700, it goes up to 1909 years before Christ, nearly, as has been said.

It must be remembered, that what is here investigated is the limit, or the most modern date possible to be assigned to the observations in question. The supposition that $x-y$ $=a$, is the most probable of all, and it gives $x=4801$, which corresponds to the beginning of the Calyougham.

on the moon's acceleration, has become more worthy of attention, and more conclusive. For that acceleration is no longer a mere empirical equation, introduced to reconcile the ancient observations with the modern, nor a fact that can only be accounted for by hypothetical causes, such as the resistance of the ether, or the time necessary for the transmission of gravity; it is a phenomenon, which Laplace has, with great ability, deduced from the principle of universal gravitation, and shown to be necessarily connected with the changes in the eccentricity of the earth's orbit, discovered by Lagrange; so that the acceleration of the moon is indirectly produced by the action of the planets, which alternately increasing and diminishing the said eccentricity, subjects the moon to different degrees of that force by which the sun disturbs the time of her revolution round the earth. It is therefore a periodical inequality, by which the moon's motion, in the course of ages, will be as much retarded as accelerated; but its changes are so slow, that her motion has been constantly accelerated, even for a longer period than that to which the observations of India extend.

A formula for computing the quantity of this inequality, has been given by Laplace, which,

---

* Mém. Acad. des Scienc. 1786, p. 235, &c.

though only an approximation, being derived from theory, is more accurate than that which Mayer deduced entirely from observation; * and if it be taken instead of Mayer's, which last, on account of its simplicity, I have employed in the preceding calculations, it will give a quantity somewhat different, though not such as to affect the general result. It makes the acceleration for 4383 years, dated from the beginning of the Calyougham, to be greater by 17′, 39″, than was found from Mayer's rule, and greater consequently by 16′, 32″, than was deduced from the tables of Chrisnabouram. It is plain that this coincidence is still near enough to leave the argument that is founded on it in possession of all its force, and to afford a strong confirmation of the accuracy of the theory, and the authenticity of the tables.

That observations made in India, when all Europe was barbarous or uninhabited, and investigations into the most subtle effects of gravitation made in Europe, near five thousand years afterwards, should thus come in mutual support of one another, is perhaps the most striking example of the progress and vicissitude of science, which the history of mankind has yet exhibited.

29. This, however, is not the only instance of the same kind that will occur, if, from examining

* Mém. Acad. des Scienc. 1786, p. 260.

the radical places and mean motions in the Indian astronomy, we proceed to consider some other of its elements, such as, the length of the year, the inequality of the sun's motion, and the obliquity of the ecliptic, and compare them with the conclusions deduced from the theory of gravity, by Lagrange. To that geometer, physical astronomy is indebted for one of the most beautiful of its discoveries, viz. That all the variations in our system are periodical; so that, though every thing, almost without exception, be subject to change, it will, after a certain interval, return to the same state in which it is at present, and leave no room for the introduction of disorder, or of any irregularity that might constantly increase. Many of these periods, however, are of vast duration. A great number of ages, for instance, must elapse before the year be again exactly of the same length, or the sun's equation of the same magnitude as at present. An astronomy, therefore, which professes to be so ancient as the Indian, ought to differ considerably from ours in many of its elements. If, indeed, these differences are irregular, they are the effects of chance, and must be accounted errors; but if they observe the laws, which theory informs us that the variations in our system do actually observe, they must be held as the most undoubted marks of authenticity. We

* Mém. de l'Acad. de Berlin, 1782, p. 170, &c.

are to examine, as Bailly has done, which of these takes place in the case before us. *

30. The tables of Tirvalore, which, as we have seen, refer their date to the beginning of the Calyougham, make the sidereal year to consist of 365 *d.* 6 *h.* 12′, 30″; and therefore the tropical of 365 *d.* 5 *h.* 50′, 35″, which is 1′, 46″ longer than that of Lacaille. † Now, the tropical year was in reality longer at that time than it is at present; for though the sidereal year, or the time which the earth takes to return from one point of space to the same point again, is always of the same magnitude, yet the tropical year being affected by the precession of the equinoxes, is variable by a small quantity, which never can exceed 3′, 40″, and which is subject to slow and unequal alternations of diminution and increase. A theorem, expressing the law and the quantity of this variation, has been investigated by Lagrange, in the excellent Memoir already mentioned; ‡ and it makes the year 3102 before Christ, 40″½ longer than the year at the beginning of the present century. § The year in the tables of Tirvalore is therefore too great by 1′, 5″½.

31. But the determination of the year is always

---

* Astr. Indienne, p. 160, &c.

† *Supra*, § 18 and 10.

‡ Mém. Acad. Berlin, 1782, p. 289.

§ Astr. Indienne, p. 160.

from a comparison of observations made at a considerable interval from one another; and, even to produce a degree of accuracy much less than what we see belongs to the tables of Tirvalore, that interval must have been of several ages. Now, says Bailly, if we suppose these observations to have been made in that period of 2400 years, immediately preceding the Calyougham, to which the Brahmins often refer; and if we also suppose the inequality of the precession of the equinoxes to increase as we go back, in proportion to the square of the times, we shall find, that, at the middle of this period, or 1200 years before the beginning of the Calyougham, the length of the year was 365 *d.* 5 *h.* 50′, 41″, almost precisely as in the tables of Tirvalore. And hence it is natural to conclude, that this determination of the solar year is as ancient as the year 1200 before the Calyougham, or 4300 before the Christian era. *

32. In this reasoning, however, it seems impossible to acquiesce; and Bailly himself does not appear to have relied on it with much confidence. †
We are not at liberty to suppose that the precession of the equinoxes increases in the ratio abovementioned, or, which is the same, that the equinoctial points

---

* Astr. Indienne, p. 161.

† He says, " Sans doute il ne peut rèsulter de ce calcul qu'un appercu."

go back with a motion equably retarded. If, by Lagrange's formula, we trace back, step by step, the variation of the solar year, we shall find, that about the beginning of the Calyougham, it had nearly attained the extreme point of one of those vibrations, which many centuries are required to complete; and that the year was then longer than it has ever been since, or than it had been for many ages before. It was 40½″ longer than it is at present; but, at the year 5500 before Christ, it was only 29″ longer than at present, instead of 2′, 50″, which is the result of Bailly's supposition. During all the intervening period of 2400 years, the variation of the year was between these two quantities; and we cannot, therefore, by any admissible supposition, reduce the error of the tables to less than 1′, 5″. The smallness of this error, though extremely favourable to the antiquity, as well as the accuracy of the Indian astronomy, is a circumstance from which a more precise conclusion can hardly be deduced.

33. The equation of the sun's centre is an element in the Indian astronomy, which has a more unequivocal appearance of belonging to an earlier period than the Calyougham. The maximum of that equation is fixed, in these tables, at 2°, 10′, 32″. It is at present, according to Lacaille, 1°, 55½′, that is, 15′ less than with the Brahmins. Now, Lagrange has shown, that the sun's equation, to-

gether with the eccentricity of the earth's orbit, on which it depends, is subject to alternate diminution and increase, and accordingly has been diminishing for many ages. In the year 3102 before our era, that equation was 2°, 6′, 28½″; less, only by 4′, than in the tables of the Brahmins. But if we suppose the Indian astronomy to be founded on observations that preceded the Calyougham, the determination of this equation will be found to be still more exact. Twelve hundred years before the commencement of that period, or about 4300 years before our era, it appears, by computing from Lagrange's formula, that the equation of the sun's centre was actually 2°, 8′, 16″; so that if the Indian astronomy be as old as that period, its error, with respect to this equation, is but of 2′.*

34. The obliquity of the ecliptic is another element in which the Indian astronomy and the European do not agree, but where their difference is exactly such as the high antiquity of the former is found to require. The Brahmins make the obliquity of the ecliptic 24°. Now, Lagrange's formula for the variation of the obliquity, † gives 22′, 32″, to be added to its obliquity in 1700, that is, to 23°, 28′, 41″, in order to have that which took place in the year 3102 before our era. This gives us

* Astr. Ind. p. 163.

† Mém. Acad. Berlin, 1782, p. 287.

23°, 51′, 13″, which is 8′, 47″, short of the determination of the Indian astronomers. But if we suppose, as in the case of the sun's equation, that the observations on which this determination is founded, were made 1200 years before the Calyougham, we shall find that the obliquity of the ecliptic was 23°, 57′, 45″, and that the error of the tables did not much exceed 2′. *

35. Thus, do the measures which the Brahmins assign to these three quantities, the length of the tropical year, the equation of the sun's centre, and the obliquity of the ecliptic, all agree in referring the epoch of their determination to the year 3102 before our era, or to a period still more ancient. This coincidence in three elements, altogether independent of one another, cannot be the effect of chance. The difference, with respect to each of them, between their astronomy and ours, might singly perhaps be ascribed to inaccuracy; but that three errors, which chance had introduced, should be all of such magnitudes, as to suit exactly the same hypothesis concerning their origin, is hardly to be conceived. Yet there is no other alternative, but to admit this very improbable supposition, or to acknowledge that the Indian astronomy is as ancient as one, or other of the periods above mentioned.

---

* Astr. Ind. p. 165.

36. This conclusion would receive great additional confirmation, could we follow Bailly in his analysis of the astronomy of the planets, contained in the tables of Chrisnabouram ;* but the length to which this *paper* is already extended, will allow only a few of the most remarkable particulars to be selected.

In these tables, which are for the epoch 1491, the mean motions are given with considerable accuracy, but without an appearance of being taken from Ptolemy, or any of the astronomers already mentioned. Two inequalities, called the *schigram* and the *manda*, are also distinguished in each of the planets, both superior and inferior.† The first of these is the same with that which we call the parallax of the earth's orbit, or the apparent inequality of a planet, which arises not from its own motion, but from that of the observer; but whether it is ascribed, in the Indian astronomy, to its true cause, or to the motion of the planet in an epicycle, is a question about which the tables give no direct information. The magnitude, however, of this equation is assigned, for each of the planets, with no small exactness, and is varied, in the different points of its orbit, by a law which approaches very near to the truth.

The other inequality coincides with that of the

* Astr. Ind. p. 173, &c. † Ibid. p. 177.

planet's centre, or that which arises from the eccentricity of its orbit, and it is given near the truth for all the planets, except Mercury, by which, as is no wonder, the first astronomers were, everywhere, greatly deceived. Of this inequality, it is supposed, just as in the cases of the sun and moon, that it is always as the sine of the planet's distance from the point of its slowest motion, or from what we call its aphelion, and is consequently greatest at 90° from that point.

It were to be wished that we knew the etymology of the names which are given to these inequalities, as it might explain the theory which guided the authors of the tables. The titles of our astronomical tables, the terms *aphelion*, *heliocentric*, or *geocentric place*, &c. would discover the leading ideas of the Copernican system, were no other description of it preserved.

37. In the manner of applying these two inequalities, to correct the mean place of a planet, the rules of this astronomy are altogether singular. In the case of a superior planet, they do not make use of the mean anomaly, as the argument for finding out the equation *manda*, but of that anomaly, when corrected first by half the equation *schigram*, and afterwards by half the equation *manda*.* By the equation of the centre, obtained with this argument,

* Astr. Ind. p. 194.

the mean longitude of the planet is corrected, and its true heliocentric place consequently found, to which there is again applied the parallax of the annual orbit, that the geocentric place may be obtained. The only difficulty here, is in the method of taking out from the tables the equation to the centre. It is evidently meant for avoiding some inaccuracy, which was apprehended from a more direct method of calculation, but of which, even after the ingenious remarks of Bailly, it seems impossible to give any clear and satisfactory account.

38. The manner of calculating the places of the inferior planets has a great resemblance to the former; with this difference, however, that the equation *manda*, or of the centre, is applied to correct, not the mean place of the planet, but the mean place of the sun; and to this last, when so corrected, is applied the equation *schigram*, which involves the planet's elongation from the sun, and gives its geocentric place.* This necessarily implies, that the centre, about which the inferior planets revolve, has the same apparent mean motion with the sun: but whether it be a point really different from the sun, or the same; and, if the same, whether it be in motion or at rest, are left entirely undetermined, and we know not, whether in the astronomy of India, we have here discovered a resemblance to the

---

* Astr. Ind. p. 199, &c.

Ptolemaic, the Tychonic, or the Copernican system.

39. These tables, though their radical places are for the year 1491 of our era, have an obvious reference to the great epoch of the Calyougham. For if we calculate the places of the planets from them, for the beginning of the astronomical year, at that epoch, we find them all in conjunction with the sun in the beginning of the moveable zodiac, their common longitude being 10^s, 6°.* According to our tables, there was, at that time, a conjunction of all the planets, except Venus, with the sun; but they were, by no means, so near to one another as the Indian astronomy represents. It is true, that the exact time of a conjunction cannot be determined by direct observation; but this does not amount to an entire vindication of the tables; and there is reason to suspect, that some superstitious notions, concerning the beginning of the Calyougham, and the signs by which nature must have distinguished so great an epoch, has, in this instance at least, perverted the astronomy of the Brahmins. There are, however, some coincidences between this part of their astronomy, and the theory of gravity, which must not be forgotten.

40. The first of these respects the aphelion of Jupiter, which, in the tables, is supposed to have a

* Astr. Ind. p. 181.

retrograde motion of 15° in 200000 years,* and to have been, at the epoch of 1491, in longitude $5^{s}, 21^{\circ}, 40', 20''$, from the beginning of the zodiac. It follows, therefore, that in the year 3102 before Christ, the longitude of Jupiter's aphelion was $3^{s}, 27^{\circ}, 0'$, reckoned from the equinox. Now, the same, computed from Lalande's tables, is only $3^{s}, 16^{\circ}, 48', 58''$; so that there would seem to be an error of more than 10° in the tables of the Brahmins. But, if it be considered, that Jupiter's orbit is subject to great disturbances, from the action of Saturn, which Lalande does not profess to have taken into account, we will be inclined to appeal once more to Lagrange's formulas, before we pass sentence against the Indian astronomy.†

From one of these formulas, we find, that the true place of the aphelion of Jupiter, at the time above mentioned, was $3^{s}, 26^{\circ}, 50', 40''$, which is but $10', 40''$, different from the tables of Chrisnabouram. The French and Indian tables are, therefore, both of them exact, and only differ because they are adapted to ages near five thousand years distant from one another.

41. The equation of Saturn's centre is an instance of the same kind. That equation, at pre-

* Astr. Ind. p. 184, § 13.

† Mém. Acad. Berlin, 1782, p. 246. Astr. Ind. p. 186.

sent, is, according to Lalande, 6°, 23′, 19″; and hence, by means of one of the formulas above mentioned, Bailly calculates, that, 3102 years before Christ, it was 7°, 41′, 22″. * The tables of the Brahmins make it 7°, 39′, 44″, which is less only by 1′, 38″, than the preceding equation, though greater than that of the present century by 1°, 16′, 25″.

42. Bailly remarks, that the equations for the other planets are not given with equal accuracy, and afford no more such instances as the former. But it is curious to observe, that new researches into the effects of gravitation, have discovered new coincidences of the same kind; and that the two great geometers, who have shared between them the glory of perfecting the *theory of disturbing forces*, have each contributed his part to establish the antiquity of the Indian astronomy. Since the publication of Bailly's work, two other instances of an exact agreement, between the elements of these tables, and the conclusions deduced from the theory of gravity, have been observed, and communicated to him by Laplace, in a letter, inserted in the *Journal des Savans.*

In seeking for the cause of the secular equations, which modern astronomers have found it necessary to apply to the mean motion of Jupiter and Saturn,

* Astr. Ind. p. 188.

Laplace has discovered, that there are inequalities belonging to both these planets, arising from their mutual action on one another, which have long periods, one of them no less than 877 years; so that the mean motion must appear different, if it be determined from observations made in different parts of those periods. "Now, I find," says he, "by my theory, that at the Indian epoch of 3102 years before Christ, the apparent and annual mean motion of Saturn was 12°, 13′, 14″, and the Indian tables make it 12°, 13′, 13″.

"In like manner, I find, that the annual and apparent mean motion of Jupiter at that epoch was 30°, 20′, 42″, precisely as in the Indian astronomy." *

43. Thus have we enumerated no less than nine astronomical elements, † to which the tables of India assign such values as do, by no means, belong to them in these later ages, but such as the theory of gravity proves to have belonged to them three thousand years before the Christian era. At that time, therefore, or in the ages preceding it, the ob-

---

* Esprit des Journaux, Nov. 1787, p. 80.

† The inequality of the precession of the equinoxes, (§ 22;) the acceleration of the moon; the length of the solar year; the equation of the sun's centre; the obliquity of the ecliptic; the place of Jupiter's aphelion; the equation of Saturn's centre; and the inequalities in the mean motion of both these planets.

servations must have been made from which these elements were deduced. For it is abundantly evident, that the Brahmins of later times, however willing they might be to adapt their tables to so remarkable an epoch as the Calyougham, could never think of doing so, by substituting, instead of quantities which they had observed, others which they had no reason to believe had ever existed. The elements in question are precisely what these astronomers must have supposed invariable, and of which, had they supposed them to change, they had no rules to go by for ascertaining the variations; since, to the discovery of these rules is required, not only all the perfection to which astronomy is, at this day, brought in Europe, but all that which the sciences of motion and of extension have likewise attained. It is no less clear, that these coincidences are not the work of accident; for it will scarcely be supposed that chance has adjusted the errors of the Indian astronomy with such singular felicity, that observers. who could not discover the true state of the heavens, at the age in which they lived, have succeeded in describing one which took place several thousand years before they were born.

44. The argument, however, which regards the originality of these tables, is, in some measure, incomplete, till we have considered the geometrical principles which have been employed in their con-

struction. For it is not impossible, that when seen connected by those principles, and united into general theorems, they may be found to have relations to the Greek astronomy, which did not appear, when the parts were examined singly. On this subject, therefore, I am now to offer a few observations.

45. The rules by which the phenomena of eclipses are deduced from the places of the sun and moon, have the most immediate reference to geometry; and of these rules, as found among the Brahmins of Tirvalore, M. Legentil has given a full account, in the Memoir that has been so often quoted. We have also an account of the method of calculation used at Chrisnabouram by Father Duchamp.*

It is a necessary preparation, in both of these, to find the time of the sun's continuance above the horizon, at the place and the day for which the calculation of an eclipse is made, and the rule by which the Brahmins resolve this problem, is extremely simple and ingenious. At the place for which they calculate, they observe the shadow of a gnomon on the day of the equinox, at noon, when the sun, as they express it, is in the middle of the world. The height of the gnomon is divided into 720 equal parts, in which parts the length of the

* Astr. Ind. p. 355, &c.

shadow is also measured. One third of this measure is the number of minutes by which the day, at the end of the first month after the equinox, exceeds twelve hours; four-fifths of this excess is the increase of the day during the second month; and one third of it is the increase of the day, during the third month. *

46. It is plain, that this rule involves the supposition, that, when the sun's declination is given, the same ratio every where exists between the arch which measures the increase of the day at any place, and the tangent of the latitude; for that tangent is the quotient which arises from dividing the length of the shadow by the height of the gnomon. Now, this is not strictly true; for such a ratio only subsists between the chord of the arch, and the tangent above mentioned. The rule is, therefore, but an approximation to the truth, as it necessarily supposes the arch in question to be so small as to coincide nearly with its chord. This supposition holds only of places in low latitudes; and the rule which is founded on it, though it may safely be applied in countries between the tropics, in those that are more remote from the equator, would lead into errors too considerable to escape observation. †

---

* Mém. Acad. des Scienc. II. p. 175.

† To judge of the accuracy of this approximation, suppose

As some of the former rules, therefore, have served to fix the time, so does this, in some measure, to ascertain the place of its invention. It is the simplification of a general rule, adapted to the circumstances of the torrid zone, and suggested to the astronomers of Hindostan by their peculiar

---

O to be the obliquity of the ecliptic, and x the excess of the semidiurnal arch, on the longest day, above an arch of 90°, then sin. x=tan.O × tan. lat. But if G be the height of a gnomon, and S the length of its shadow on the equinoctial day, $\frac{S}{G}$= tan. lat. and sin. x= tan.O × $\frac{S}{G}$. Therefore x= tan.O × $\frac{S}{G}+\frac{\tan.O^3 \times S^3}{6G^3}+\frac{\tan.O^5 \times S^5}{24G^5}+$ &c. or in minutes of time, reckoned after the Indian manner, x=572.957 (tan.O × $\frac{S}{G}$+ tan.O^3 × $\frac{S^3}{6G^3}$+ &c.)

If O=24°, then tan O=.4452, and the first term of this formula gives $x=572.957 \times \frac{.4452S}{G}=\frac{255S}{G}$, which is the same with the rule of the Brahmins.

For that rule, reduced into a formula, is $2x=\frac{720S}{G}\left(\frac{1}{3}+\frac{4}{15}+\frac{1}{9}\right)=\frac{512S}{G}$, or $x=\frac{256S}{G}$.

They have therefore computed the coefficient of $\frac{S}{G}$ with sufficient accuracy; the error produced by the omission of the rest of the terms of the series will not exceed 1′, even at the tropics, but, beyond them, it increases fast, and, in the latitude of 45°, would amount to 8′.

situation. It implies the knowledge of the circles of the sphere, and of spherical trigonometry, and perhaps argues a greater progress in mathematical reasoning than a theorem that was perfectly accurate would have done. The first geometers must naturally have dreaded nothing so much as any abatement in the rigour of their demonstrations, because they would see no limits to the error and uncertainty in which they might, by that means, be involved. It was long before the mathematicians of Greece understood how to set bounds to such errors, and to ascertain their utmost extent, whether on the side of excess or defect; in this art they appear to have received the first lessons so late as the age of Archimedes.

47. The Brahmins having thus obtained the variations of the length of the day, at any place, or what we call the ascensional differences, apply them likewise to another purpose. As they find it necessary to know the point of the ecliptic, which is on the horizon, at the time when an eclipse happens, they have calculated a table of the right ascensions of the points of the ecliptic in time, to which they apply the ascensional differences for the place in question, in order to have the time which each of the signs takes to descend below the horizon of that place.* This is exactly the method,

* Acad. des Scienc. 1772, II. p. 205.

as is well known, which the most skilful astronomer, in like circumstances, would pursue. Their table of the differences of right ascension is but for a few points in the ecliptic, viz. the beginning of each sign, and is only carried to minutes of time, or tenths of a degree. It is calculated, however, so far as it goes, with perfect accuracy, and it supposes the obliquity of the ecliptic, as before, to be twenty-four degrees.

Such calculations could not be made without spherical trigonometry, or some method equivalent to it. If, indeed, we would allow the least skill possible to the authors of these tables, we may suppose that the arches were measured on the circles of a large globe, or armillary sphere, such as we know to have been one of the first instruments of the Egyptian and Greek astronomers. But there are some of the tables where the arches are put down true to seconds, a degree of accuracy which a mechanical method can scarcely have afforded.

48. In another part of the calculation of eclipses, a direct application is made of one of the most remarkable propositions in geometry. In order to have the semiduration of a solar eclipse, they subtract from the square of the sum of the semidiameters of the sun and moon, the square of a certain line, which is a perpendicular from the centre of the sun on the path of the moon; and from the remainder, they extract the square root, which is the measure

of the semiduration.* The same thing is practised in lunar eclipses.† These operations are all founded on a very distinct conception of what happens in the case of an eclipse, and on the knowledge of this theorem, that, in a right-angled triangle, the square on the hypothenuse is equal to the squares on the other two sides. It is curious to find the theorem of Pythagoras in India, where, for aught we know, it may have been discovered, and from whence that philosopher may have derived some of the solid, as well as the visionary speculations, with which he delighted to instruct or amuse his disciples.

49. We have mentioned the use that is made of the semidiameters of the sun and moon in these calculations, and the method of ascertaining them, is deserving of attention. For the sun's apparent diameter, they take four-ninths of his diurnal motion, and for the moon's diameter, one twenty-fifth of her diurnal motion. In an eclipse, they suppose the section of the shadow of the earth, at the distance of the moon, to have a diameter five times that of the moon; and in all this, there is considerable accuracy, as well as great simplicity. The apparent diameters of the sun and moon, increase and diminish with their angular velocities; and though there be a mistake in supposing, that they do so ex-

* Mém. Acad. des Scienc. 1772, II. p. 259.

† Ibid. 241.

actly in the same proportion, it is one which, without telescopes and micrometers, cannot easily be observed. The section of the earth's shadow, likewise, if the sun's apparent diameter be given, increases as the moon's increases, or as her distance from the earth diminishes, and nearly enough in the same ratio to justify the rule which is here laid down.

50. The historian of the Academy of Sciences, in giving an account of M. Legentil's Memoir, has justly observed, that the rule described in it, for finding the difference between the true and apparent conjunction, at the time of a solar eclipse, contains the calculation of the moon's parallax, but substitutes the parallax in right ascension for the parallax in longitude ;* an error which the authors of this astronomy would probably have avoided, had they derived their knowledge from the writings of Ptolemy. From this supposed parallax in longitude, they next derive the parallax in latitude, where we may observe an application of the doctrine of similar triangles; for they suppose the first of these to be to the last in the constant ratio of 25 to 2, or nearly as the radius to the tangent of the inclination of the moon's orbit to the plane of the ecliptic. We have here, therefore, the application of another geometrical theorem, and that too

* Hist. Acad. II. p. 109. Ibid. Mém. 253—256.

proceeding on the supposition, that a small portion of the sphere, on each side of the point which the sun occupies at the middle of the eclipse, may be held to coincide with a plane touching it in that point.

51. The result which the Brahmins thus obtain will be allowed to have great accuracy, if it be considered how simple their rules are, and how long it must be since their tables were corrected by observations. In two eclipses of the moon, calculated in India by their method, and likewise observed there by M. Legentil, the error, in neither case, exceeded 23′ of time, (corresponding to one of 13′ of a degree, in the place of the moon;) and in the duration and magnitude of the eclipse, their calculation came still nearer to the truth.*

* In the language, however, of their rules, we may trace some marks of a fabulous and ignorant age, from which indeed even the astronomy of Europe is not altogether free. The place of the moon's ascending node, is with them *the place of the Dragon* or *the Serpent;* the moon's distance from the node, is literally translated by M. Legentil *la lune offensée du dragon.* Whether it be that we have borrowed these absurdities from India, along with astrology, or if the popular theory of eclipses has, at first, been every where the same, the moon's node is also known with us by the name of the *cauda draconis.* In general, however, the signification of the terms in these rules, so far as we know it, is more rational. In one of them we may remark considerable refinement; *ayanangsam,* which is the name for the reduction

52. Since an inequality was first observed in the motions of the sun or moon, the discovery of the law which it follows, and the method of determining the quantity of it, in the different points of their orbits, has been a problem of the greatest importance; and it is curious to inquire, in what manner the astronomers of India have proceeded to resolve it. For this purpose, we must examine the tables of the *chaiaa*, or equations of the centre for the sun and moon, and of the *manda*, or equations of the centre for the planets. With respect to the first, as contained in the tables of Siam, M. Cassini observed, that the equations followed the ratio of the sines of the mean distances from the apogee; but as they were calculated only for a few points of the orbit, it could not be known with what degree of exactness this law was observed. Here, however, the tables of Chrisnabouram remove the uncertainty, as they give the equation of the centre for every degree of the mean motion, and make it nearly as the sine of the distance from the apogee.

They do so, however, only nearly; and it will be found on trial, that there is, in the numbers of

---

made on the sun's longitude, on account of the precession of the equinoxes, is compounded from *ayanam*, a *course*, and *angsam*, an *atom*. Mém. Acad. II. p. 251. The equinox is almost the only point not distinguished by a visible objects of which the *course* or motion is computed in this astronomy.

the table, a small, but regular variation from this law, which is greatest when the argument is 30°, though even there it does not amount to a minute. The sun's equation, for instance, which, when greatest, or when the argument is 90°, is, by these tables, 2°, 10′, 32″, should be, when the argument is 30°, just the half of this, or 1°, 5′, 16″, did the numbers in the table follow exactly the ratio of the sines of the argument. It is, however, 1°, 6′, 3″; and this excess of 47″ cannot have arisen from any mistake about the ratio of the sine of 30° to that of 90°, which is shown to be that of 1 to 2, by a proposition in geometry* much too simple to have been unknown to the authors of these tables. The rule, therefore, of the equations, being proportional exactly to the sines of the argument, is not what was followed, or intended to be followed, in the calculation of them. The differences, also, between the numbers computed by that rule, and those in the tables, are perfectly regular, decreasing from the point of 30°, both ways toward the beginning and end of the quadrant, where they vanish altogether.

These observations apply also to the tables of Narsapur,† and to the moon's equations, as well as to the sun's, with a circumstance, however, which

* Euc. Lib. IV. Prop. 15.

† See these tables, Astr. Ind. p. 414.

is not easily accounted for, viz. that the differences between the numbers calculated by Cassini's rule, and those in the tables, are not greater in the case of the moon than of the sun, though the equation of the latter be more than double that of the former. They apply also to the tables *manda* of the planets, where the equations are greater than the ratio of the sines of their arguments requires, the excess being greatest at 30°, and amounting to some minutes in the equations of Saturn, Jupiter, and Mars, in which last it is greatest of all.

53. Though, for these reasons, it is plain, that the rule of Cassini is not the same with that of the Brahmins, it certainly includes the greater part of it; and if the latter, whatever it may have been, were expressed in a series, according to the methods of the modern analysis, the former would be the first term of that series. We are not, however, much advanced in our inquiry in consequence of this remark; for the first terms of all the series, which can, on any hypothesis, express the relation of the equation of the centre to the anomaly of a planet, are so far the same, that they are proportional to the sine of that anomaly; and it becomes therefore necessary to search among these hypotheses, for that by which the series of small differences, described above, may be best represented. It is needless to enter here into any detail of the reasonings by which this has been done, and by

which I have found, that the argument in the table bears very nearly the same relation to the corresponding numbers, that the anomaly of the eccen tric does to the equation of the centre. By the anomaly of the eccentric, however, I do not mean the angle which is known by that name in the solution of Kepler's problem, but that which serves the same purpose with it, on the supposition of a circular orbit, and an uniform angular motion about a point which is not the centre of that orbit, but which is as distant from it, on the one side, as the earth (or the place of the observer) is on the other. It is the angle, which, in such an orbit, the line drawn from the planet to the centre, makes with the line drawn from thence to the apogee; and the argument in the Indian tables coincides with this angle.

This hypothesis of a double eccentricity, is certainly not the simplest that may be formed with respect to the motion of the heavenly bodies, and is not what one would expect to meet with here; but it agrees so well with the tables, and gives the equations from the arguments so nearly, especially for the moon and the planets, that little doubt remains of its being the real hypothesis on which these tables were constructed. *

---

* The formula deduced from this hypothesis, for calculating the equation of the centre from the anomaly of the

54. Of this, the method employed to calculate the place of any of the five planets from these tables, affords a confirmation. But, in reasoning about that method, it is necessary to put out of the question the use that is made of the parallax of the annual orbit, or of the *schigram*, in order to have the argument for finding the equation of the centre, which is evidently faulty, as it makes that equation to be affected by a quantity, (the parallax of the annual orbit,) on which it has in reality no dependence. To have the rule free from error, it is to be taken, therefore, in the case when there is no parallax of the annual orbit, that is, when the planets are in opposition or conjunction with the sun. In that case, the mean anomaly is first corrected by the subtraction or addition of half the equation that belongs to it in the table. It then becomes the true argument for finding, from that same table, the equation of the centre, which is next applied to the mean anomaly, to have the true. Now, this agrees perfectly with the conclusion above; for the mean anomaly, by the subtraction or addition of half the equation belonging to it in the

---

eccentric, is the following: Let x be the equation of the centre, φ the anomaly of the eccentric, e the eccentricity of the orbit, or the tangent of half the greatest equation; then $x=2e\,\sin.\varphi+\frac{2e^3\,\sin.3\varphi}{3}+\frac{2e^5\,\sin.5\varphi}{5}+$ &c.

table, is converted, almost precisely, into the anomaly of the eccentric, and becomes therefore the proper argument for finding out the equation, which is to change the mean anomaly into the true.* There can be no doubt, of consequence, that the conclusion we have come to is strictly applicable to the planets, and that the orbit of each of them, in this astronomy, is supposed to be a circle, the earth not being in its centre, but the angular velocity of the planet being uniform about a certain point, as far from that centre on the one side, as the earth is on the opposite.

55. Between the structure of the tables of the equations of the sun and moon, and the rules for using them, there is not the same consistency; for in both of them, the argument, which we have found to be the eccentric anomaly, is nevertheless

---

* This method of calculation is so nearly exact, that even in the orbit of Mars, the equation calculated from the mean anomaly, rigorously on the principle of his angular motion being uniform, about a point distant from the centre, as described above, will rarely differ a minute from that which is taken out from the Indian tables by this rule. It was remarked, (§ 37,) that it is not easy to explain the rules for finding the argument of the equation of the centre, for the planets. What is said here explains fully one part of that rule, viz. the correction made by half the equation *manda;* the principle on which the other part proceeds, viz. the correction by half the equation *schigram,* is still uncertain.

treated as the mean. So far as concerns the sun, this leads to nothing irreconcilable with our supposition, because the sun's equation being small, the difference will be inconsiderable, whether the argument of that equation be treated as the eccentric or the mean anomaly.

But it is otherwise with respect to the moon, where the difference between considering the argument of the equation as the mean, or as the eccentric anomaly, is not insensible. The authority of the precepts, and of the tables, are here opposed to one another; and we can decide in favour of the latter, only because it leads to a more accurate determination of the moon's place than the former. It would indeed be an improvement on their method of calculation, which the Brahmins might make consistently with the principles of their own astronomy, to extend to the moon their rule for finding the equation of the centre for the planets. They would then avoid the palpable error of making the maximum of the moon's equation at the time when her mean anomaly is 90°, and would ascertain her place every where with greater exactness. It is probable that this is the method which they were originally directed to follow.

56. From the hypothesis which is thus found to be the basis of the Indian astronomy, one of the first conclusions which presents itself, is the existence of a remarkable affinity between the system of

the Brahmins and that of Ptolemy. In the latter, the same thing was supposed for the five planets, that appears in the former to have been universally established, viz. that their orbits were circles, having the earth within them, but removed at a small distance from the centre, and that each planet described the circumference of its orbit, not with an uniform velocity, but with one that would appear uniform, if it were viewed from a point as far above the centre of the orbit, as that centre is above the earth. This point was, in the language of Ptolemy's astronomy, the centre of the *Equant.*

Now, concerning this coincidence, it is the more difficult to judge, as, on the one hand, it cannot be ascribed to accident, and, on the other, it may be doubted, whether it arises necessarily out of the nature of the subject, or is a consequence of some unknown communication between the astronomers of India and of Greece.

The first hypothesis by which men endeavoured to explain the phenomena of the celestial motions, was that of a uniform motion in a circle, which had the earth for its centre. This hypothesis was, however, of no longer continuance than till instruments of tolerable exactness were directed to the heavens. It was then immediately discovered, that the earth was not the centre of this uniform motion; and the earth was therefore supposed to be placed at a certain distance from the centre of the orbit, while the planet revolved in the circumference of it with the same velocity as before. Both these steps may be accounted necessary; and in however many places of the earth, and however cut off from mutual intercourse, astronomy had begun to be cultivated, I have no doubt that these two suppositions would have succeeded one another, just as they did among the Greek astronomers.

But when more accurate observations had shown the insufficiency even of this second hypothesis, what ought naturally to be the third, may be thought not quite so obvious; and if the Greeks made choice of that which has been described above, it may seem to have been owing to certain metaphysical notions concerning the simplicity and perfection of a circular and uniform motion, which inclined them to recede from that supposition, no farther than appearances rendered absolutely necessary. The same coincidence between the ideas of metaphysics and astronomy, cannot be supposed to have taken place in other countries; and therefore, where we find this third hypothesis to have prevailed, we may conclude that it was borrowed from the Greeks.

57. Though it cannot be denied, that, in this reasoning, there is some weight, yet it must be observed, that the introduction of the third hypothesis did not rest among the Greeks altogether on the coincidence above mentioned. It was one suited to

their progress in mathematical knowledge, and offered almost the only system, after the two former were exploded, which rendered the planetary motions the subject of geometrical reasoning, to men little versed in the methods of approximation. This was the circumstance then, which, more than any other, probably influenced them in the choice of this hypothesis, though we are not to look for it as an argument stated in their works, but may judge of the influence it had, from the frequency with which, many ages afterwards, the αγεωμετρησια of Kepler's system was objected to him by his adversaries; an objection to which that great man seemed to pay more attention than it deserved.

There is reason, therefore, to think, that in every country where astronomy and geometry had neither of them advanced beyond a certain point, the hypothesis of the equant would succeed to that of a simple eccentric orbit, and therefore cannot be admitted as a proof, that the different systems in which it makes a part, are necessarily derived from the same source. Some other circumstances attending this hypothesis, as it is found in the Indian tables, go still farther, and seem quite inconsistent with the supposition that the authors of these tables derived it from the astronomers of the west. For, *first*, It is applied by them to all the heavenly bodies, that is, to the sun and moon, as well as the planets. With Ptolemy, and with all those who founded their systems on his, it extended only to the latter, insomuch that Kepler's great reformation in astronomy, the discovery of the elliptic orbits, began from his proving, that the hypothesis of the equant was as necessary to be introduced for the sake of the sun's orbit, as for those of the planets, and that the eccentricity in both cases, must be bisected. It is, therefore, on a principle no way different from this of Kepler, that the tables of the sun's motion are computed in the Indian astronomy, though it must be allowed, that the method of using them is not perfectly consistent with this idea of their construction.

2dly, The use made of the anomaly of the eccentric in these tables, as the argument of the equation of the centre, is altogether peculiar to the Indian astronomy. Ptolemy's tables of that equation for the planets, though they proceed on the same hypothesis, are arranged in a manner entirely different, and have for their argument the mean anomaly. The angle which we call the anomaly of the eccentric, and which is of so much use in the Indian tables, is not employed at all in the construction of his,* nor, I believe, in those of any other astronomer till the time of Kepler; and even by Kepler it was not made the argument of the equation to the centre. The method, explained above,

* Almagest. Lib. XI. cap. ix. and x.

of converting the mean anomaly into that of the eccentric, and consequently into the argument of the equation, is another peculiarity, and though simple and ingenious, has not the accuracy suited to the genius of the Greek astronomy, which never admitted even of the best approximation, when a rigorous solution could be found; and, on the whole, if the resemblance of these two systems, even with all the exceptions that have been stated, must still be ascribed to some communication between the authors of them, that communication is more likely to have gone from India to Greece, than in the opposite direction. It may perhaps be thought to favour this last opinion, that Ptolemy has no where demonstrated the necessity of assigning a double eccentricity to the orbits of the planets, and has left room to suspect, that authority, more than argument, has influenced this part of his system.

58. In the tables of the planets, we remarked another equation, (*schigram,*) answering to the parallax of the earth's orbit, or the difference between the heliocentric and the geocentric place of the planet. This parallax, if we conceive a triangle to be formed by lines drawn from the sun to the earth and to the planet, and also from the planet to the earth, is the angle of that triangle, subtended by the line drawn from the sun to the earth. And so, accordingly, it is computed in these tables ; for if we resolve such a triangle as is here described,

we will find the angle, subtended by the earth's distance from the sun, coincide very nearly with the schigram.

The argument of this equation is the difference between the mean longitude of the sun and of the planet. The orbits are supposed circular; but whether the inequality in question was understood to arise from the motion of the earth, or from the motion of the planet in an epicycle, the centre of which revolves in a circle, is left undetermined, as both hypotheses may be so adjusted as to give the same result with respect to this inequality. The proportional distances of the planets from the earth or the sun, may be deduced from the tables of these equations, and are not far from the truth.

59. The preceding calculations must have required the assistance of many subsidiary tables, of which no trace has yet been found in India. Besides many other geometrical propositions, some of them also involve the ratio, which the diameter of a circle was supposed to bear to its circumference, but which we would find it impossible to discover from them exactly, on account of the small quantities that may have been neglected in their calculations. Fortunately, we can arrive at this knowledge, which is very material when the progress of geometry is to be estimated, from a passage in the *Ayeen Akbery*, where we are told, that the Hindoos suppose the diameter of a circle to be to its

circumference as 1250 to 3927,* and where the author, who knew that this was more accurate than the proportion of Archimedes, (7 to 22,) and believed it to be perfectly exact, expresses his astonishment, that among so simple a people, there should be found a truth, which, among the wisest and most learned nations, had been sought for in vain.

The proportion of 1250 to 3927 is indeed a near approach to the quadrature of the circle; it differs little from that of Metius, 113 to 355, and is the same with one equally remarkable, that of 1 to 3.1416. When found in the simplest and most elementary way, it requires a polygon of 768 sides to be inscribed in a circle; an operation which cannot be arithmetically performed without the knowledge of some very curious properties of that curve, and, at least, nine extractions of the square root, each as far as ten places of decimals. All this must have been accomplished in India; for it is to be observed, that the above mentioned proportion cannot have been received from the mathematicians of the west. The Greeks left nothing on this subject more accurate than the theorem of Archimedes; and the Arabian mathematicians seem not to have attempted any nearer approximation. The geometry of modern Europe can much less be regarded

* Ayeen Akbery, Vol. III. p. 32.

as the source of this knowledge. Metius and Vieta were the first, who, in the quadrature of the circle, surpassed the accuracy of Archimedes; and they flourished at the very time when the Institutes of Akbar were collected in India.

60. On the grounds which have now been explained, the following general conclusions appear to be established.

I. The observations on which the astronomy of India is founded, were made more than three thousand years before the Christian era; and, in particular, the places of the sun and moon, at the beginning of the Calyougham, were determined by actual observation.

This follows from the exact agreement of the radical places in the tables of Tirvalore, with those deduced for the same epoch from the tables of Lacaille and Mayer, and especially in the case of the moon, when regard is had to her acceleration. It follows, too, from the position of the fixed stars in respect of the equinox, as represented in the Indian zodiac; from the length of the solar year; and, lastly, from the position and form of the orbits of Jupiter and Saturn, as well as their mean motions; in all of which, the tables of the Brahmins, compared with ours, give the quantity of the change that has taken place, just equal to that which the action of the planets on one another may be shown to have produced, in the space of forty-eight centuries, reckoning back from the beginning of the present.

Two other of the elements of this astronomy, the equation of the sun's centre, and the obliquity of the ecliptic, when compared with those of the present time, seem to point to a period still more remote, and to fix the origin of this astronomy 1000 or 1200 years earlier, that is, 4300 years before the Christian era; and the time necessary to have brought the arts of calculating and observing to such perfection as they must have attained at the beginning of the Calyougham, comes in support of the same conclusion.

Of such high antiquity, therefore, must we suppose the origin of this astronomy, unless we can believe, that all the coincidences which have been enumerated, are but the effects of chance; or, what indeed were still more wonderful, that, some ages ago, there had arisen a Newton among the Brahmins, to discover that universal principle which connects, not only the most distant regions of space, but the most remote periods of duration; and a Lagrange, to trace, through the immensity of both, its most subtle and complicated operations.

II. Though the astronomy which is now in the hands of the Brahmins is so ancient in its origin, yet it contains many rules and tables that are of later construction.

The first operation for computing the moon's place from the tables of Tirvalore, requires that 1,600,984 days should be subtracted from the time that has elapsed since the beginning of the Calyougham, which brings down the date of the rule to the year 1282 of our era. At this time, too, the place of the moon, and of her apogee, are determined with so much exactness, that it must have been done by observation, either at the instant referred to, or a few days before or after it. At this time, therefore, it is certain, that astronomical observations were made in India, and that the Brahmins were not, as they are now, without any knowledge of the principles on which their rules are founded. When that knowledge was lost, will not perhaps be easily ascertained; but there are, I think, no circumstances in the tables from which we can certainly infer the existence of it at a later period than what has just been mentioned; for, though there are more modern epochs to be found in them, they are such as may have been derived from the most ancient of all, by help of the mean motions in the tables of Chrisnabouram,* without any other skill than is required to an ordinary calculation. Of these epochs, beside what have been occasionally mentioned in the course of our remarks, there is one (involved in the tables of

---

* Astr. Indienne, p. 307.

Narsapur) as late as the year 1656, and another as early as the year 78 of our era, which marks the death of Salivaganam, one of their princes, in whose reign a reform is said to have taken place in the methods of their astronomy. There is no reference to any intermediate date, from that time to the beginning of the Calyougham.

The parts of this astronomy, therefore, are not all of the same antiquity; nor can we judge, merely from the epoch to which the tables refer, of the age to which they were originally adapted. We have seen, that the tables of Chrisnabouram, though they profess to be no older than the year 1491 of our era, are, in reality, more ancient than the tables of Tirvalore, which are dated from the Calyougham, or at least have undergone fewer alterations. This we concluded from the slow motion given to the moon, in the former of these tables, which agreed, with such wonderful precision, with the secular equation applied to that planet by Mayer, and explained by Laplace.

But it appears, that neither the tables of Tirvalore or Chrisnabouram, nor any with which we are yet acquainted, are the most ancient to be found in India. The Brahmins constantly refer to an astronomy at Benares, which they emphatically style *the ancient,* * and which they say is not now un-

---

* Astr. Indienne, p. 309. M. Legentil, Mém. Acad. Scienc. 1772. P. II. p. 221.

derstood by them, though they believe it to be much more accurate than that by which they calculate. That it is more accurate, is improbable; that it may be more ancient, no one who has duly attended to the foregoing facts and reasonings will think impossible; and every one, I believe, will acknowledge, that no greater service could be rendered to the learned world, than to rescue this precious fragment from obscurity. If that is ever to be expected, it is when the zeal for knowledge has formed a literary society among our countrymen in Bengal, and while that society is directed by the learning and abilities of Sir William Jones. Indeed, the farther discoveries which may be made with respect to this science, do not interest merely the astronomer and the mathematician, but every one who delights to mark the progress of mankind, or is curious to look back on the ancient inhabitants of the globe. It is through the medium of astronomy alone that a few rays from those distant objects can be conveyed in safety to the eye of a modern observer, so as to afford him a light, which, though it be scanty, is pure and unbroken, and free from the false colourings of vanity and superstition.

III. The basis of the four systems of astronomical tables which we have examined is evidently the same.

Though these tables are scattered over an extensive country, they seem to have been all originally adapted, either to the same meridian, or to meridians at no great distance, which traverse what we may call the classical ground of India, marked by the ruins of Canoge, Palibothra, and Benares. They contain rules that have originated between the tropics; whatever be their epoch, they are all, by their mean motions, connected with that of the Calyougham; and they have besides one uniform character which it is perhaps not easy to describe. Great ingenuity has been exerted to simplify their rules; yet, in no instance almost, are they reduced to the utmost simplicity; and when it happens that the operations to which they lead are extremely obvious, these are often involved in an artificial obscurity. A Brahmin frequently multiplies by a greater number than is necessary, where he seems to gain nothing but the trouble of dividing by one that is greater in the same proportion; and he calculates the era of Salivaganam with the formality of as many distinct operations, as if he were going to determine the moon's motion since the beginning of the Calyougham. The same spirit of exclusion, the same fear of communicating his knowledge, seems to direct the calculus which pervades the religion of the Brahmin; and in neither of them is he willing to receive or to impart instruction. With all these circumstances of resemblance, the methods of this astronomy are as much diversified as we can suppose the same system to be, by passing through the hands of a succession of ingenious men, fertile in resources, and acquainted with the variety and extent of the science which they cultivated. A system of knowledge, which is thus assimilated to the genius of the people, that is diffused so widely among them, and diversified so much, has a right to be regarded, either as a native, or a very ancient inhabitant of the country where it is found.

IV. The construction of these tables implies a great knowledge of geometry, arithmetic, and even of the theoretical part of astronomy.

In proof of this, it is unnecessary to recapitulate the remarks that have been already made. It may be proper, however, to add, that the method of calculating eclipses, to which these tables are subservient, is, in no respect, an empirical one, founded on the mere observation of the intervals at which eclipses return, one after another, in the same order. It is indeed remarkable, that we find no trace here of the period of 6585 days and 8 hours, or 223 lunations, the *Saros* of the Chaldean astronomers, which they employed for the prediction of eclipses, and which (observed with more or less accuracy) the first astronomers every where must have employed, before they were able to analyse eclipses, and to find out the laws of every cause contributing to them. That empirical method, if it once existed in India, is now forgotten, and has long since given place to the more scientific and accurate one, which offers a complete analysis of the phenomena, and calculates, one by one, the motions of the sun, of the moon, and of the node.

But what, without doubt, is to be accounted the greatest refinement in this system, is the hypothesis employed in calculating the equations of the centre for the sun, moon, and planets, that, viz. of a circular orbit having a double eccentricity, or having its centre in the middle, between the earth and the point about which the angular motion is uniform.* If to this we add the great extent of

---

* It should have been remarked before, that Bailly has taken notice of the analogy between the Indian method of calculating the places of the planets, and Ptolemy's hypothesis of the equant, though on different principles from those that have been followed here, and such as do not lead to the same conclusion. In treating of the question, whether the sun or earth has been supposed the centre of the planetary motions by the authors of this astronomy, he says, "Ils semblent avoir reconnu que les deux inégalités (l'équation du centre et la parallaxe de l'orbe annuel) etoient vues de deux centres differens; et dans l'impossibilité où ils étoient de déterminer et le lieu et la distance des deux centres, ils ont imaginé de rapporter les deux inégalités à un point qui tînt le milieu, c'est-à-dire, à un point également éloigné du soleil, et de la terre. Ce nouveau centre ressemble assez au centre de l'équant de Ptolemée. (Astr. Ind. Disc. Prél. p. 69.) The fictitious centre, which Bailly compares with the equant of Ptolemy, is therefore a point which bisects the

geometrical knowledge requisite to combine this, and the other principles of their astronomy toge-

---

distance between the sun and earth, and which, in some respects, is quite different from that equant; the fictitious centre, which, in the preceding remarks, is compared with the equant of Ptolemy, is a point of which the distance from the earth is bisected by the centre of the orbit, precisely as in the case of that equant. Bailly draws his conclusion from the use made of half the equation *schigram*, as well as half the equation *mânda*, in order to find the argument of this last equation. The conclusion here is established, by abstracting altogether from the former, and considering the cases of oppositions and conjunctions, when the latter equation only takes place. If, however, the hypothesis of the equant shall be found of importance in the explanation of the Indian astronomy, it must be allowed that it was first suggested by Bailly, though in a sense very different from what it is understood in here, and from what it was understood in by Ptolemy.

For what farther relates to the parts of the astronomy of Chaldea and of Greece, which may be supposed borrowed from that of India, I must refer to the 10th Chap. of the *Astronomie Indienne,* where that subject is treated with great learning and ingenuity. After all, the silence of the ancients with respect to the Indian astronomy, is not easily accounted for. The first mention that is made of it, is by the Arabian writers; and Bailly quotes a very singular passage, where Massoudi, an author of the twelfth century, says that Brama composed a book, entitled, *Sind-Hind,* that is, *Of the Age of Ages,* from which was composed the book *Maghisti,* and from thence the *Almagest* of Ptolemy. (Astr. Ind. Disc. Prél. p. 175.)

ther, and to deduce from them the just conclusions; the possession of a calculus equivalent to trigonometry; and, lastly, their approximation to the quadrature of the circle, we shall be astonished at the magnitude of that body of science, which must have enlightened the inhabitants of India in some remote age, and which, whatever it may have communicated to the western nations, appears to have received nothing from them.

Such are the conclusions that seem to me to follow, with the highest probability, from the facts which have been stated. They are, without doubt, extraordinary; and have no other claim to our belief, except that, as I think has been fully proved, their being false were much more wonderful than their being true. There are but few things, however, of which the contrary is impossible. It must

---

The fabulous air of this passage is, in some measure, removed, by comparing it with one from Abulfaragius, who says that, under the celebrated Almaimon, the 7th Khalif of Babylon, (about the year 813 of our era,) the astronomer Habash composed three sets of astronomical tables, one of which was *ad regulas Sind-Hind;* that is, as Mr Costard explains it, according to the rules of some Indian treatise of astronomy. (Asiatic Miscel. Vol. I. p. 34.) The *Sind-Hind* is therefore the name of an astronomical book that existed in India in the time of Habash, and the same, no doubt, which Massoudi says was ascribed to Brama.

be remembered, that the whole evidence on this subject is not yet before the public, and that the repositaries of Benares may contain what is to confirm or to invalidate these observations.

FINIS.

ON

THE ORIGIN

AND

INVESTIGATION OF PORISMS.

ON

THE ORIGIN

AND

INVESTIGATION OF PORISMS. *

1. THE restoration of the ancient books of geometry would have been impossible, without the coincidence of two circumstances, of which, though the one is purely accidental, the other is essentially connected with the nature of the mathematical sciences. The first of these circumstances is the preservation of a short abstract of those books, drawn up by Pappus Alexandrinus, together with a series of such *lemmata* as he judged useful to facilitate the study of them. The second is, the necessary connection that takes place among the objects of every mathematical work, which, by excluding whatever is arbitrary, makes it possible to determine the whole course of an investigation, when only a few points in it are known. From the union of these circumstances, mathematics has

* From the Transactions of the Royal Society of Edinburgh, Vol. III. (1794.)—ED.

enjoyed an advantage of which no other branch of knowledge can partake; and while the critic or the historian has only been able to lament the fate of those books of Livy and Tacitus which are lost, the geometer has had the high satisfaction to behold the works of Euclid and Apollonius reviving under his hands.

2. The first restorers of the ancient books were not, however, aware of the full extent of the work which they had undertaken. They thought it sufficient to demonstrate the propositions, which they knew from Pappus, to have been contained in those books; but they did not follow the ancient method of investigation, and few of them appear to have had any idea of the elegant and simple analysis by which these propositions were originally discovered, and by which the Greek geometry was peculiarly distinguished.

Among these few, Fermat and Halley are to be particularly remarked. The former, one of the greatest mathematicians of the last age, and a man in all respects of superior abilities, had very just notions of the geometrical analysis, and appears often abundantly skilful in the use of it; yet in his restoration of the Loci Plani, it is remarkable, that in the most difficult propositions, he lays aside the analytical method, and contents himself with giving the synthetical demonstration. The latter, among the great number and variety of his literary occupations, found time for a most attentive study of the ancient mathematicians, and was an instance of, what experience shows to be much rarer than might be expected, a man equally well acquainted with the ancient and the modern geometry, and equally disposed to do justice to the merit of both. He restored the books of Apollonius, on the problem De Sectione Spatii, according to the true principles of the ancient analysis.

These books, however, are but short, so that the first restoration of considerable extent that can be reckoned complete, is that of the Loci Plani by Dr Simson, published in 1749, which, if it differs at all from the work it is intended to replace, seems to do so only by its greater excellence. This much at least is certain, that the method of the ancient geometers does not appear to greater advantage in the most entire of their writings, than in the restoration above mentioned; and that Dr Simson has often sacrificed the elegance to which his own analysis would have led, in order to tread more exactly in what the *lemmata* of Pappus pointed out to him, as the track which Apollonius had pursued.

3. There was another subject, that of Porisms, the most intricate and enigmatical of any thing in the ancient geometry, which was still reserved to exercise the genius of Dr Simson, and to call forth that enthusiastic admiration of antiquity, and that

unwearied perseverance in research, for which he was so peculiarly distinguished. A treatise in three books, which Euclid had composed on Porisms, was lost, and all that remained concerning them was an abstract of that treatise, inserted by Pappus Alexandrinus, in his Mathematical Collections, in which, had it been entire, the geometers of later times would doubtless have found wherewithal to console themselves for the loss of the original work. But unfortunately it has suffered so much from the injuries of time, that all which we can *immediately* learn from it is, that the ancients put a high value on the propositions which they called Porisms, and regarded them as a very important part of their analysis. The Porisms of Euclid are there said to be, "Collectio artificiosissima multarum rerum quæ spectant ad analysin difficiliorum et generalium problematum." * The curiosity, however, which is excited by this encomium, is quickly disappointed; for when Pappus proceeds to explain what a Porism is, he lays down two definitions of it, one of which is rejected by him as imperfect, while the other, which is stated as correct, is too vague and indefinite to convey any useful information.

These defects might nevertheless have been supplied, if the enumeration which he next gives of

* Collectiones Math. Lib. VII. in init.

Euclid's propositions had been entire; but on account of the extreme brevity of his enunciations, and their reference to a diagram which is lost, and for the constructing of which no directions are given, they are all, except one, perfectly unintelligible. For these reasons, the fragment in question is so obscure, that even to the learning and penetration of Dr Halley, it seemed impossible that it could ever be explained; and he therefore concluded, after giving the Greek text with all possible correctness, and adding the Latin translation, "Hactenus Porismatum descriptio nec mihi intellecta, nec lectori profutura. Neque aliter fieri potuit, tam ob defectum schematis cujus fit mentio, quam ob omissa quædam et transposita, vel aliter vitiata in propositionis generalis expositione, unde quid sibi velit Pappus haud mihi datum est conjicere. His adde dictionis modum nimis contractum, ac in re difficili, qualis hæc est, minime usurpandum." *

4. It is true, however, that before this time, Fermat had attempted to explain the nature of Porisms, and not altogether without success. † Guiding his conjectures by the definition which Pappus censures as imperfect, because it defined

---

* De Sectione Rationis, proem. p. 37.

† "Porismatum Euclidæorum renovata doctrina, et sub forma Isagoges exhibita." Fermat Opera Varia, p. 116.

Porisms only "ab accidente," viz. "Porisma est quod deficit hypothesi a Theoremate Locali," he formed to himself a tolerably just notion of these propositions, and illustrated his general description by examples that are in effect Porisms. But he was able to proceed no farther; and he neither proved that his notion of a Porism was the same with Euclid's, nor attempted to restore, or explain any one of Euclid's propositions, much less did he suppose that they were to be investigated by an analysis peculiar to themselves. And so imperfect, indeed, was this attempt, that the complete restoration of the Porisms was necessary to prove that Fermat had even approximated to the truth.

5. All this did not, however, deter Dr Simson from turning his thoughts to the same subject, which he appears to have done very early, and long before the publication of the Loci Plani in 1749. The account he gives of his progress, and of the obstacles he encountered, will be always interesting to mathematicians. "Postquam vero apud Pappum legeram Porismata Euclidis collectionem fuisse artificiosissimam multarum rerum, quæ spectant ad analysin difficiliorum et generalium problematum, magno desiderio tenebar, aliquid de iis cognoscendi; quare sæpius et multis variisque viis tum Pappi propositionem generalem, mancam et imperfectam, tum primum lib. i. Porisma, quod solum ex omnibus in tribus libris integrum adhuc manet, intelli-

gere et restituere conabar; frustra tamen, nihil enim proficiebam. Cumque cogitationes de hac re multum mihi temporis consumpserint, atque molestæ admodum evaserint, firmiter animum induxi hæc nunquam in posterum investigare; præsertim cum optimus geometra Halleius spem omnem de iis intelligendis abjecisset. Unde quoties menti occurrebant, toties eas arcebam. Postea tamen accidit, ut improvidum et propositi immemorem invaserint, meque detinuerint donec tandem lux quædam effulserit, quæ spem mihi faciebat inveniendi saltem Pappi propositionem generalem, quam quidem multa investigatione tandem restitui. Hæc autem paulo post una cum Porismate primo lib. i. impressa est inter Transactiones Phil. anni 1723, No. 177." *

The propositions here mentioned, as inserted in the Philosophical Transactions for 1723, are all that Dr Simson published on the subject of Porisms during his life, though he continued his investigations concerning them, and succeeded in restoring a great number of Euclid's propositions, together with their analysis. The propositions thus restored form a part of that valuable edition of the posthumous works of this geometer which the mathematical world owes to the munificence of the late Earl Stanhope.

6. The subject of Porisms is not, however, ex-

* Rob. Simson, Op. reliqua, p. 319.

hausted, nor is it yet placed in so clear a light as to need no farther illustration. It yet remains to inquire into the probable origin of these propositions, that is to say, into the steps by which the ancient geometers appear to have been led to the discovery of them. It remains also to point out the relations in which they stand to the other classes of geometrical truths; to consider the species of analysis, whether geometrical or algebraical, that belongs to them; and, if possible, to assign the reason why they have so long escaped the notice of modern mathematicians. It is to these points that the following observations are chiefly directed.

I begin with describing the steps that appear to have led the ancient geometers to the discovery of Porisms; and must here supply the want of express testimony by probable reasonings, such as are necessary, whenever we would trace remote discoveries to their sources, and which have more weight in mathematics than in any other of the sciences.

7. It cannot be doubted, that it has been the solution of problems which, in all states of the mathematical sciences, has led to the discovery of most geometrical truths. The first mathematical inquiries, in particular, must have occurred in the form of questions, where something was given, and something required to be done; and by the reasonings necessary to answer these questions, or to

discover the relation between the things that were given, and those that were to be found, many truths were suggested, which came afterwards to be the subjects of separate demonstration. The number of these was the greater, that the ancient geometers always undertook the solution of problems with a scrupulous and minute attention, which would scarcely suffer any of the collateral truths to escape their observation. We know from the examples which they have left us, that they never considered a problem as resolved, till they had distinguished all its varieties, and evolved separately every different case that could occur, carefully remarking whatever change might arise in the construction, from any change that was supposed to take place among the magnitudes which were given.

Now, as this cautious method of proceeding was not better calculated to avoid error than to lay hold of every truth that was connected with the main object of inquiry, these geometers soon observed, that there were many problems which, in certain circumstances, would admit of no solution whatever, and that the general construction by which they were resolved would fail, in consequence of a particular relation being supposed among the quantities which were given. Such problems were then said to become impossible; and it was readily perceived, that this always happened when one of the conditions prescribed was inconsistent with the rest, so that the supposition of their being united in the same *subject* involved a contradiction. Thus, when it was required to divide a given line, so that the rectangle under its segments should be equal to a given space, it was evident that, if this space was greater than the square of half the given line, the thing required could not possibly be done; the two conditions, the one defining the magnitude of the line, and the other that of the rectangle under its segments, being then inconsistent with one another. Hence an infinity of beautiful propositions concerning the *maxima* and the *minima* of quantities, or the limits of the possible relations which quantities may stand in to one another.

8. Such cases as these would occur even in the solution of the simplest problems; but when geometers proceeded to the analysis of such as were more complicated, they must have remarked, that their constructions would sometimes fail, for a reason directly contrary to that which has now been assigned. Instances would be found where the lines that, by their intersection, were to determine the thing sought, instead of intersecting one another, as they did in general, or of not meeting at all, as in the above mentioned case of impossibility, would coincide with one another entirely, and leave the question of consequence unresolved. But though this circumstance must have created considerable embarrassment to the geometers who first

observed it, as being, perhaps, the only instance in which the language of their own science had yet appeared to them ambiguous or obscure, it would not probably be long till they found out the true interpretation to be put on it. After a little reflection, they would conclude, that, since, in the general problem, the magnitude required was determined by the intersection of the two lines above mentioned, that is to say, by the points common to them both; so, in the case of their coincidence, as all their points were in common, every one of these points must afford a solution; which solutions, therefore, must be infinite in number; and also, though infinite in number, they must all be related to one another, and to the things given, by certain laws, which the position of the two coinciding lines must necessarily determine.

On inquiring farther into the peculiarity in the state of the data which had produced this unexpected result, it might likewise be remarked, that the whole proceeded from one of the conditions of the problem involving another, or necessarily including it; so that they both together made, in fact, but one, and did not leave a sufficient number of independent conditions, to confine the problem to a single solution, or to any determinate number of solutions. It was not difficult afterwards to perceive, that these cases of problems formed very curious propositions, of an intermediate nature between problems and theorems, and that they admitted of being enunciated separately, in a manner peculiarly elegant and concise. It was to such propositions, so enunciated, that the ancient geometers gave the name of *Porisms.*

9. This deduction requires to be illustrated by examples. Suppose, therefore, that it is proposed to resolve the following problem :

Prop. I. Prob. Fig. 6.

A circle ABC, a straight line DE, and a point F, being given in position, to find a point G in the straight line DE, such that GF, the line drawn from it to the given point, shall be equal to GB, the line drawn from it touching the given circle.

Suppose the point G to be found, and GB to be drawn touching the circle ABC in B; let H be the centre of the circle ABC; join HB, and let HD be perpendicular to DE; from D draw DL, touching the circle ABC in L, and join HL. Also from the centre G, with the distance GB or GF, describe the circle BKF, meeting HD in the points K and K′.

It is plain, that the lines HD and DL are given in position and in magnitude. Also, because GB touches the circle ABC, HBG is a right angle;

and since G is the centre of the circle BKF, therefore HB touches the circle BKF, and consequently the square of HB or of HL, is equal to the rectangle K′HK. But the rectangle K′HK, together with the square of DK, is equal to the square of DH, because KK′ is bisected in D; therefore the squares of HL and DK are also equal to the square of DH. But the squares of HL and LD are equal to the same square of DH; wherefore the square of DK is equal to the square of DL, and the line DK to the line DL. But DL is given in magnitude; therefore DK is given in magnitude, and K is therefore a given point. For the same reason, K′ is a given point, and the point F being also given by hypothesis, the circle BKF is given in position. The point G therefore, the centre of the circle BKF is given, which was to be found.

Hence this construction: Having drawn HD perpendicular to DE, and DL touching the circle ABC, make DK and DK each equal to DL, and find G the centre of a circle described through the points K, F and K′; that is, let FK′ be joined, and bisected at right angles by the line MN, which meets DE in G; G will be the point required, or it will be such a point, that if GB be drawn from it, touching the circle ABC, and GF to the given point, GB and GF will be equal to one another.*

* This solution of the problem was suggested to me by

The synthetical demonstration needs not be added; but it is necessary to remark, that there are cases in which this construction fails altogether.

For, first, if the given point F be any where in the line HD, as at F′, it is evident, that MN becomes parallel to DE, and that the point G is no where to be found, or, in other words, is at an infinite distance from D.

This is true in general; but if the given point F coincide with K, then the line MN evidently coincides with DE; so that, agreeably to a remark already made, every point of the line DE may be taken for G, and will satisfy the conditions of the problem; that is to say, GB will be equal to GK, wherever the point G be taken in the line DE. The same is true if F coincide with K′.

This is easily demonstrated synthetically; for if G be any point whatsoever in the line DE, from which GB is drawn touching the circle ABC; if DK and DK′ be each made equal to DL; and if a circle be described through the points B, K, and K′; then, since the rectangle KHK′, together with the square of DK, that is, of DL, is equal to the square of DH, that is, to the squares of DL and LH, the rectangle KHK is equal to the square of HB, so that HB touches the circle BKK′. But

Professor Robison; and is more simple than that which I had originally given.

BG is at right angles to HB; therefore the centre of the circle BKK is in the line BG; and it is also in the line DE; therefore G is the centre of the circle BKK′, and GB is equal to GK.

Thus we have an instance of a problem, and that too a very simple one, which is in general determinate, admitting only of one solution, but which nevertheless, in one particular case, where a certain relation takes place among the things given, becomes indefinite, and admits of innumerable solutions. The proposition which results from this case of the problem is a Porism, according to the remarks that were made above, and in effect will be found to coincide with the 66th proposition in Dr Simson's Restoration. It may be thus enunciated: "A circle ABC being given in position, and also a straight line DE, which does not cut the circle, a point K may be found, such, that if G be any point whatever in the line DE, the straight line drawn from G to the point K, shall be equal to the straight line drawn from G, touching the circle ABC."

10. The following Porism is also derived in the same manner from the solution of a very simple problem:

PROP. II. PROB. FIG. 7.

A triangle ABC being given, and also a point D, to draw through D a straight line DG, such, that perpendiculars being drawn to it from the three angles of the triangle, viz. AE, BG, CF, the sum of the two perpendiculars on the same side of DG, shall be equal to the remaining perpendicular; or, that AE and BG together, may be equal to CF.

Suppose it done: Bisect AB in H, join CH, and draw HK perpendicular to DG.

Because AB is bisected in H, the two perpendiculars AE and BG are together double of HK; and as they are also equal to CF by hypothesis, CF must be double of HK, and CL of LH. Now, CH is given in position and magnitude; therefore the point L is given; and the point D being also given, the line DL is given in position, which was to be found.

The construction is obvious. Bisect AB in H, join CH, and take HL equal to one third of CH; the straight line which joins the points D and L is the line required.

Now, it is plain, that while the triangle ABC remains the same, the point L also remains the same, wherever the point D may be. The point D may therefore coincide with L; and when this happens, the position of the line to be drawn is left undetermined; that is to say, any line whatever drawn through L will satisfy the conditions of the problem.

Here therefore we have another indefinite case of a problem, and of consequence another Porism, which may be thus enunciated: "A triangle being given in position, a point in it may be found, such, that any straight line whatever being drawn through that point, the perpendiculars drawn to this straight line from the two angles of the triangle which are on one side of it, will be together equal to the perpendicular that is drawn to the same line from the angle on the other side of it."

11. This Porism may be made much more general; for if, instead of the angles of a triangle, we suppose ever so many points to be given in a plane, a point may be found, such, that any straight line being drawn through it, the sum of all the perpendiculars that fall on that line from the given points on one side of it, is equal to the sum of the perpendiculars that fall on it from all the points on the other side of it.

Or still more generally, any number of points being given not in the same plane, a point may be found, through which if any plane be supposed to pass, the sum of all the perpendiculars which fall on that plane from the points on one side of it, is equal to the sum of all the perpendiculars that fall on the same plane from the points on the other side of it.

It is unnecessary to observe, that the point to be found in these propositions, is no other than the centre of gravity of the given points; and that therefore we have here an example of a Porism very well known to the modern geometers, though not distinguished by them from other theorems.

12. The problem which follows appears to have led to the discovery of more than one Porism.

Prop. III. Prob. Fig. 8.

A circle ABC, and two points D and E, in a diameter of it being given, to find a point F in the circumference of the given circle, from which, if straight lines be drawn to the given points E and D, these straight lines shall have to one another the given ratio of α to β.*

Suppose the problem resolved, and that F is found, so that FE has to FD the given ratio of α to β. Produce EF any how to B, bisect the angle EFD by the line FL, and the angle DFB by the line FM.

* The ratio of α to β is supposed that of a greater to a less.

Then, because the angle EFD is bisected by FL, EL is to LD as EF to FD, that is, in a given ratio; and as ED is given, each of the segments EL, LD, is given, and also the point L.

Again, because the angle DFB is bisected by FM, EM is to MD as EF to FD, that is, in a given ratio; and therefore, since ED is given, EM, MD, are also given, and likewise the point M.

But because the angle LFD is half of the angle EFD, and the angle DFM half of the angle DFB, the two angles LFD, DFM, are equal to the half of two right angles, that is, to a right angle. The angle LFM being therefore a right angle, and the points L and M being given, the point F is in the circumference of a circle described on the diameter LM, and consequently given in position.

Now, the point F is also in the circumference of the given circle ABC; it is therefore in the intersection of two given circumferences, and therefore is found.

Hence this construction: Divide ED in L, so that EL may be to LD in the given ratio of α to β; and produce ED also to M, so that EM may be to MD in the same given ratio of α to β. Bisect LM in N, and from the centre N, with the distance NL, describe the semicircle LFM, and the point F, in which it intersects the circle ABC, is the point required, or that from which FE and FD are to be drawn.

The synthetical demonstration follows so readily from the preceding analysis, that it is not necessary to be added.

It must however be remarked, that the construction fails when the circle LFM falls either wholly without, or wholly within the circle ABC, so that the circumferences do not intersect; and in these cases the solution is impossible. It is plain also, that in another case the construction will fail, viz. when it so happens that the circumference LFM wholly coincides with the circumference ABC. In this case, it is farther evident, that every point in the circumference ABC will answer the conditions of the problem, which therefore admits of innumerable solutions, and may, as in the foregoing instances, be converted into a Porism.

13. We are therefore to inquire, in what circumstances the point L may coincide with the point A, and the point M with the point C, and of consequence the circumference LFM with the circumference ABC.

On the supposition that they coincide, EA is to AD, and also EC to CD, as α to β; and therefore EA is to EC as AD to CD, or, by conversion, EA to AC as AD to the excess of CD above AD, or to twice DO, O being the centre of the circle ABC. Therefore also, EA is to AO, or the half of AC, as AD to DO, and EA together with AO, to AO, as AD together with DO, to DO; that is,

EO to AO as AO to DO, and so the rectangle EO.OD equal to the square of AO.

Hence, if the situation of the given points E and D, (fig. 9,) in respect of the circle ABC, be such, that the rectangle EO.OD is equal to the square of AO, the semidiameter of the circle; and if, at the same time, the given ratio of α to β be the same with that of EA to AD, or of EC to CD, the problem admits of innumerable solutions; and as it is manifest, that if the circle ABC, and one of the points D or E be given, the other point, and also the ratio which is required to render the problem indefinite, may be found, therefore we have this Porism: "A circle ABC being given, and also a point D, a point E may be found, such, that the two lines inflected from these points to any point whatever in the circumference ABC, shall have to one another a given ratio, which ratio is also to be found."

This Porism is the second in the treatise De Porismatibus, where Dr Simson gives it, not as one of Euclid's propositions, but as an illustration of his own definition. It answers equally well for the purpose I have here in view, the explaining the origin of Porisms; and I have been the more willing to introduce it, that it has afforded me an opportunity of giving what seems to be the simplest investigation of the second proposition in the second book of the Loci Plani, by proving, as has been done above, that on the hypothesis of that proposition, LFM (fig. 8,) is a right angle, and L and M given points.

14. Hence also an example of the derivation of Porisms from one another. For the circle ABC, and the points E and D, remaining as in the last construction, (fig. 9,) if through D we draw any line whatever HDB, meeting the circle in B and H, and if the lines EB, EH be also drawn, these lines will cut off equal circumferences BF and HG. Let FC be drawn, and it is plain from the foregoing analysis, that the angles DFC, CFB are equal. Therefore if OG, OB be drawn, the angles BOC, COG are equal, and consequently the angles DOB, DOG. In the same manner, by joining AB, the angle DBE being bisected by BA, it is evident, that the angle AOF is equal to the angle AOH, and therefore the angle FOB to the angle HOG, that is, the arch FB to the arch HG.

Now, it is plain, that if the circle ABC, and one of the points D or E be given, the other point may be found; therefore we have this Porism, which appears to have been the last but one in the third book of Euclid's Porisms.* "A point being given, either without or within a circle given in position, if there be drawn, any how through that point, a line cutting the circle in two points; ano-

---

* Simson De Porismatibus, Prop. 53.

ther point may be found, such, that if two lines be drawn from it to the points, in which the line already drawn cuts the circle, these two lines will cut off from the circle equal circumferences."

There are other Porisms that may be deduced from the same original problem, (§ 12,) all connected, as many remarkable properties of the circle are, with the harmonical division of the diameter.

15. The preceding proposition also affords a good illustration of the general remark that was made above, concerning the conditions of a problem being involved in one another, in the Porismatic, or indefinite case. Thus, several independent conditions are here laid down, by help of which the problem is to be resolved: Two points D and E are given, (fig. 8,) from which two lines are to be inflected, and a circumference ABC, in which these lines are to meet, as also a ratio, which they are to have to one another.* Now, these conditions are all independent of each other, so that any one of them may be changed, without any change whatever in the rest. This at least is true in general; but nevertheless in one case, viz. when the given points are so related to one another, that the rectangle under their distances from the centre, is equal to the square of the radius of the circle, it

* The given points, and the centre of the given circle, are understood, throughout, to be in the same straight line.

follows from the foregoing analysis, that the ratio which the inflected lines are to have to one another, is no longer a matter of choice, but is a necessary consequence of this disposition of the points. For if any other ratio were now assigned than that of AO to OD, or, which is the same, of EA to AD, it would easily be shown, that no lines having that ratio could be inflected from the points E and D, to any point in the circle ABC. Two of the conditions are therefore reduced into one; and hence it is that the problem is indefinite.

16. From this account of the origin of Porisms, it follows, that a Porism may be defined, *A proposition affirming the possibility of finding such conditions as will render a certain problem indeterminate, or capable of innumerable solutions.*

To this definition, the different characters which Pappus has given will apply without difficulty. The propositions described in it, like those which he mentions, are, strictly speaking, neither theorems nor problems, but of an intermediate nature between both; for they neither simply enunciate a truth to be demonstrated, nor propose a question to be resolved; but are affirmations of a truth, in which the determination of an unknown quantity is involved. In as far, therefore, as they assert, that a certain problem may become indeterminate, they are of the nature of theorems; and in as far as they seek to discover the conditions by which that is brought about, they are of the nature of problems.

17. In the preceding definition, also, and the instances from which it is deduced, we may trace that imperfect description of Porisms which Pappus ascribes to the later geometers, viz. "Porisma est quod deficit hypothesi a theoremate locali." Now, to understand this, it must be observed, that if we take the converse of one of the propositions called Loci, and make the construction of the figure a part of the hypothesis, we have what was called by the ancients a Local Theorem. And again, if, in enunciating this theorem, that part of the hypothesis which contains the construction be suppressed, the proposition arising from thence will be a Porism; for it will enunciate a truth, and will also require, to the full understanding and investigation of that truth, that something should be found, viz. the circumstances in the construction, supposed to be omitted.

Thus, when we say; If from two given points E and D, (fig. 9,) two lines EF and FD are inflected to a third point F, so as to be to one another in a given ratio, the point F is in the circumference of a circle given in position: we have a Locus.

But when conversely it is said; If a circle ABC, of which the centre is O, be given in position, as also a point E, and if D be taken in the line EO, so that the rectangle EO OD be equal to the square of AO, the semidiameter of the circle; and if from E and D, the lines EF and DF be inflected to any point whatever in the circumference ABC; the ratio of EF to DF will be a given ratio, and the same with that of EA to AD: we have a local theorem.

And, lastly, when it is said; If a circle ABC be given in position, and also a point E, a point D may be found, such, that if the two lines EF and FD be inflected from E and D to any point whatever F, in the circumference, these lines shall have a given ratio to one another: the proposition becomes a Porism, and is the same that has been just investigated.

Here it is evident, that the local theorem is changed into a Porism, by leaving out what relates to the determination of the point D, and of the given ratio. But though all propositions formed in this way, from the conversion of Loci, be Porisms, yet all Porisms are not formed from the conversion of Loci. The first and second of the preceding, for instance, cannot by conversion be changed into Loci; and therefore the definition which describes all Porisms as being so convertible, is not sufficiently comprehensive. Fermat's idea of Porisms, as has been already observed, was founded wholly on this definition, and therefore could not fail to be imperfect.

18. It appears, therefore, that the definition of

Porisms given above, (§ 16,) agrees with Pappus's idea of these propositions, as far at least as can be collected from the imperfect fragment which contains his general description of them. It agrees, also, with Dr Simson's definition, which is this: * "Porisma est propositio in qua proponitur demonstrare rem aliquam, vel plures datas esse, cui, vel quibus, ut et cuilibet ex rebus innumeris, non quidem datis, sed quæ ad ea quæ data sunt eandem habent relationem, convenire ostendendum est affectionem quandam communem in propositione descriptam."

It cannot be denied, that there is a considerable degree of obscurity in this definition; † notwith-

* Simson's Opera Reliqua, p. 323.

† The following translation will perhaps be found to remedy some of the obscurity complained of.

"A Porism is a proposition, in which it is proposed to demonstrate, that one or more things are given, between which and every one of innumerable other things, not given, but assumed according to a given law, a certain relation, described in the proposition, is to be shown to take place."

It may be proper to remark, that there is an ambiguity in the word *given*, as used here and on many other occasions, where it denotes indifferently, things that are both *determinate* and *known*, and things that, though *determinate*, are *unknown*, provided they can be found. This holds as to the first application of the term in the above definition; from which, however, no inconveniency arises, when the reader is apprised of it. In the course of this paper, I have endeavoured, as much as possible, to avoid the like ambiguity.

standing of which, it is certain, that every proposition to which it applies must contain a *problematical* part, viz. " in qua proponitur demonstrare rem aliquam, vel plures datas esse ;" and also a *theoretical* part, which contains the property, or *communis affectio*, affirmed of certain things which have been previously described.

It is also evident, that the subject of every such proposition is the relation between magnitudes of three different kinds ; determinate magnitudes, which are given ; determinate magnitudes, which are to be found ; and indeterminate magnitudes, which, though unlimited in number, are connected with the others by some common property. Now, these are exactly the conditions contained in the definition that has been given here.

19. To confirm the truth of this theory of the origin of Porisms, or at least the justness of the notions founded on it, I must add a quotation from an Essay on the same subject, by a member of this Society, the extent and correctness of whose views make every coincidence with his opinions peculiarly flattering. In a paper read several years ago, before the Philosophical Society, Professor Dugald Stewart defined a Porism to be, " A proposition affirming the possibility of finding one or more of the conditions of an indeterminate theorem ;" where, by an indeterminate theorem, as he had previously explained it, is meant one which expresses a rela-

tion between certain quantities that are determinate and certain others that are indeterminate, both in magnitude and in number. The near agreement of this with the definition and explanations which have been given above, is too obvious to require to be pointed out; and I have only to observe, that it was not long after the publication of Simson's posthumous works, when, being both of us occupied in speculations concerning Porisms, we were led separately to the conclusions which I have now stated. *

* In an inquiry into the origin of Porisms, the etymology of the term ought not to be forgotten. The question, indeed, is not about the derivation of the word Πορισμα, for concerning that there is no doubt; but about the reason why this term was applied to the class of propositions above described. Two opinions may be formed on this subject, and each of them with considerable probability.

1*mo*, One of the significations of πορίζω, is *to acquire* or *obtain*; and hence Πορισμα, *the thing obtained* or *gained*. Accordingly, Scapula says, "Est vox a geometris desumpta qui theorema aliquid ex demonstrativo syllogismo necessario sequens inferentes, illud quasi *lucrari* dicuntur, quod non ex professo quidem theorematis hujus instituta sit demonstratio, sed tamen ex demonstratis recte sequatur." In this sense, Euclid uses the word in his Elements of Geometry, where he calls the corollaries of his propositions, *Porismata*. This circumstance creates a presumption, that when the word was applied to a particular class of propositions, it was meant, in both cases, to convey nearly the same idea, as it is not at all probable, that so correct a writer as Euclid, and so scrupu-

20. We might next proceed to consider the particular Porisms which Dr Simson has restored, and

---

lous in his use of words, should employ the same term to express two ideas which are perfectly different. May we not therefore conjecture, that these propositions got the name of Porisms, entirely with a reference to their origin? According to the idea explained above, they would in general occur to mathematicians when engaged in the solution of the more difficult problems, and would arise from those particular cases, where one of the conditions of the data involved in it some one of the rest. Thus, a particular kind of theorem would be obtained, following as a corollary from the solution of the problem; and to this theorem the term Πορισμα might be very properly applied, since, in the words of Scapula, already quoted, "Non ex professo theorematis hujus instituta sit demonstratio, sed tamen ex demonstratis recte sequatur."

2do, But though this interpretation agrees so well with the supposed origin of Porisms, it is not free from difficulty. The verb πορίζω has another signification, to find out, to discover, to devise; and is used in this sense by Pappus, when he says, that the propositions called Porisms, afford great delight, τοις δυναμενοις ορᾶν και πορίζειν, to those who are able to understand and *investigate*. Hence comes πορισμος, the act of finding out, or discovering, and from πορισμος, in this sense, the same author evidently considers Πορισμα as being derived. His words are, Εφασαν δε (όι αρχαιοι) Πορισμα ειναι το προτεινομενον εις Πορισμον αυτȣ τȣ προτεινομενȣ, the ancients said, that a Porism is something proposed for the *finding out*, or *discovering* of the very thing proposed. It seems singular, however, that Porisms should have taken their name from a circumstance common to them with so many other geome-

to show, that every one of them is the indeterminate case of some problem. But of this it is so easy for any one, who has attended to the preceding remarks, to satisfy himself, by barely examining the enunciations of those propositions, that the detail into which it would lead seems to be unnecessary. I shall therefore go on to make some observations on that kind of *analysis* which is particularly adapted to the investigation of Porisms.

If the idea which we have given of these propositions be just, it follows, that they are always to be discovered, by considering the cases in which the construction of a problem fails, in consequence of the lines which, by their intersection, or the points which, by their position, were to determine the magnitude required, happening to coincide with one another. A Porism may therefore be deduced from the problem it belongs to, in the same manner that the propositions concerning the *maxima* and *minima* of quantities are deduced from the problems of which they form the limitations; and such

---

trical truths; and if this was really the case, it must have been on account of the enigmatical form of their enunciation, which required, that in the analysis of these propositions, a sort of double discovery should be made, not only of the *truth*, but also of the *meaning* of the very thing which was proposed. They may therefore have been called Porismata or Investigations, by way of eminence.

no doubt is the most natural and most obvious analysis of which this class of propositions will admit.

It is not, however, the only one that they will admit of; and there are good reasons for wishing to be provided with another, by means of which, a Porism that is any how suspected to exist, may be found out, independently of the general solution of the problem to which it belongs. Of these reasons, one is, that the Porism may perhaps admit of being investigated more easily than the general problem admits of being resolved; and another is, that the former, in almost every case, helps to discover the simplest and most elegant solution that can be given of the latter.

The truth of this last observation has been already exemplified in two of the preceding problems, where the Porismatic case, by determining the point K in the first, and L in the second of them, became necessary to the general solution. In more difficult problems, the same will be found to hold still more remarkably, and this is evidently what Pappus had in view, when, in a passage already quoted, he called Porisms, " Collectio artificiosissima multarum rerum quæ spectant ad analysin difficiliorum et generalium problematum."

On this account, it is desirable to have a method of investigating Porisms, which does not require, that we should have previously resolved the problems they are connected with, and which may always serve to determine, whether to any given problem there be attached a Porism, or not. Dr Simson's analysis may be considered as answering to this description; for as that geometer did not regard these propositions at all in the light that is done here, nor in relation to their origin, an independent analysis of this kind, was the only one that could occur to him; and he has accordingly given one which is extremely ingenious, and by no means easy to be invented, but which he uses with great skilfulness and dexterity throughout the whole of his Restoration.

It is not easy to ascertain whether this be the precise method used by the ancients. Dr Simson had here nothing to direct him but his genius, and has the full merit of the first inventor. It seems probable, however, that there is at least a great affinity between the methods, since the *lemmata* given by Pappus as necessary to Euclid's demonstrations, are subservient also to those of our modern geometer.

21. I shall employ the same sort of analysis in the Porisms that follow, at least till we come to treat of them algebraically, where a method of investigating these propositions will present itself, which is perhaps more simple and direct than any other. The following Porism is the first of Euclid's, and the first also that was restored. It is given here to exemplify the advantage which, in investigations

of this kind, may be derived from employing the law of continuity in its utmost extent, and pursuing Porisms to those extreme cases, where the indeterminate magnitudes increase *ad infinitum;* into which state Dr Simson probably did not think it safe to follow them, and was thereby deprived of no inconsiderable help toward the simplifying of his constructions. If therefore it can be shown, that this help may be obtained without any sacrifice of geometrical accuracy, it will be some improvement in this branch of the analysis.

The Porism just mentioned may be considered as having occurred in the solution of a problem. Suppose it were required; two points A and B, (fig. 10,) and also three straight lines DE, FK, KL, being given in position, together with two points H and M, in two of these lines, to inflect from A and B to a point in the third line, two lines that shall cut off from KF and KL two segments, adjacent to the given points H and M, having to one another the given ratio of α to β.

Now, in order to find whether there be any Porism connected with this problem, suppose that there is, and that the following proposition is true.

Prop. IV. Porism. Fig. 10.

22. Two points A and B, and two straight lines DE and FK, being given in position, and also a

point H in one of them, a line LK may be found, and also a point in it M, both given in position, such, that AE and BE, inflected from the points A and B to any point whatsoever of the line DE, shall cut off from the other lines FK and LK, segments, HG and MN, adjacent to the given points H and M, having to one another the given ratio of α to β.

First, let AE′, BE′ be inflected to the point E′, so that AE′ may be parallel to FK, then shall E′B be parallel to KL, the line to be found. For if it be not parallel to KL, the point of their intersection must be at a finite distance from the point M, and therefore making as β to α, so this distance to a fourth proportional, the distance from H, at which AE′ intersects FK, will be equal to that fourth proportional. But AE′ does not intersect FK, for they are parallel by construction; therefore BE′ cannot intersect KL; KL is therefore parallel to BE′, a line given in position.

Again, let AE″, BE″ be inflected to E″, so that AE″ may pass through the given point H; then it is plain, that BE″ must pass through the point to be found M; for if not, it may be demonstrated, just as has been done above, that AE″ does not pass through H, contrary to the supposition. The point to be found is therefore in the line E″B, which is given in position.

Now, if from E there be drawn EP parallel to AE′, and ES parallel to BE′, BS is to SE as BL to LN, and AP to PE as AF to FG; wherefore the ratio of FG to LN is compounded of the ratios of AF to BL, PE to SE, and BS to AP. But the ratio of PE to SE is the same with that of AE′ to BE′, and the ratio of BS to AP is the same with that of DB to DA, because DB is to BS as DE′ to E′E, or as DA to AP. Therefore the ratio of FG to LN is compounded of the ratios of AF to BL, AE′ to BE′, and DB to DA.

In like manner, because E″ is a point in the line DE, and AE″, BE″ are inflected to it, the ratio of FH to LM, is compounded of the same ratios of AF to BL, AE′ to BE′, and DB to DA; and therefore the ratio of FH to LM is the same with that of FG to NL, and the same consequently with that of HG to MN. But the ratio of HG to MN is given, being by supposition that of α to β; the ratio of FH to LM is therefore also given, and FH being given, LM is given in magnitude. Now, LM is parallel to BE′, a line given in position; therefore M is in a line QM parallel to AB, and given in position. But the point M is also in another line BE″ given in position; therefore the point M, and also the line KML drawn through it parallel to BE′, are given in position, which were to be found.

The construction is thus: From A draw AE′

parallel to FK, meeting DE in E′; join BE′, and take in it BQ, so that as α to β so HF to BQ, and through Q draw QM parallel to AB. Let HA be drawn, and produced till it meet DE in E″, and let BE″ be drawn meeting QM in M. Through M draw KML parallel to BE′; then is KML the line, and M the point, which were to be found.

It is plain, that there are two lines which will answer the conditions of the Porism; for if in QB, produced on the other side of B, there be taken Bq equal to BQ; and if qm be drawn parallel to AB, intersecting MB in m; and if $m\lambda$ be drawn parallel to BQ, the part mn, cut off by EB produced, will be equal to MN, and have to HG the ratio required.

It is plain also, that whatever be the ratio of α to β, and whatever be the magnitude of FH, if the other things given remain the same, the lines found will be all parallel to BE′. But if the ratio of α to β remain the same likewise, and if only the point H vary, the position of KL will remain the same, and the point M will vary.

23. This construction, from which, and the foregoing analysis, the synthetical demonstration follows readily, will be found to be more simple than Dr Simson's, owing entirely to the use that has been made of the law of continuity in the two extreme cases, where, according to the language of the modern analysis, HG becomes infinite, in the

one, and equal to nothing, in the other. Had it been affirmed, agreeably to that same language, that in the first of those cases, because of the constant ratio of HG to MN, these lines must both become infinite at the same time, and in the second, that for the same reason they must both vanish at the same time, we might have been accused of departing from the strict form of reasoning employed in the ancient geometry. But when the thing is stated as above, and it is proved, that when AE′ does not meet KF, it is impossible for BE′ to meet ML; and again, that when AE″ passes through H, it is impossible for BE″ not to pass through M, the air of paradox is entirely removed, and the tracing of the law of continuity is rendered perfectly consistent with the utmost severity of geometrical demonstration.

Dr Simson has applied this Porism very ingeniously to the solution of the same problem from which it is here supposed to be derived; * and it is worthy of remark, that supposing the points A and B, and the lines DE and FK to be as in the figure of this Porism, if the third of the given lines be not parallel to BE′, that problem can always be resolved, and admits of two solutions; but if it be parallel to BE′, the problem either becomes impossible, or a Porism; that is, it either admits of no solution, or

* Opera Reliqua, de Porismatibus, Prop. 25.

of an infinite number. We shall soon have occasion to extend the same observation to other Porisms.

Another general remark which I have to make on the analysis of Porisms is, that it frequently happens, as in the last example, that the magnitudes required may all, or a part of them, be found by considering the extreme cases; but for the discovery of the relation between them, and the indefinite magnitudes, or *res innumeræ*, we must have recourse to the hypothesis of the Porism in its most general, or indefinite form, and must endeavour so to conduct the reasoning, that the indefinite magnitudes shall at length wholly disappear, and leave a proposition, containing only a relation of determinate magnitudes to one another. Now, in order to accomplish this, Dr Simson frequently employs two statements of the general hypothesis, which he compares together; as for instance, in his analysis of the last Porism, he assumes, not only E, any point whatsoever in the line DE, but also another point O, any whatsoever in the same line, to both of which he supposes lines to be inflected from the given points A and B. This double statement, however, cannot be made, without rendering the investigation long and complicated; and therefore it may be of use to remark, that it is never necessary, but may always be avoided by an appeal to simpler Porisms, or to Loci, or to the propositions of the Data. I shall give the following Porism as an example, where this is done with some difficulty, but with considerable advantage, in regard to the simplicity and shortness of the investigation.

Prop. V. Porism. Fig. 11.

24. Let there be three straight lines AB, AC, CB given in position, and from any point whatsoever in one of them, as D, let perpendiculars be drawn to the other two, as DF, DE; a point G may be found, such, that if GD be drawn from it to the point D, the square of that line shall have a given ratio to the sum of the squares of the perpendiculars DF and DE, which ratio is to be found.

Draw from A and B the lines AH, BK at right angles to BC and CA, and divide AB in L, so that AL may be to LB in the given ratio of the square of AH to the square of BK, or, which is the same, of the square of AC to the square of CB. The point L is therefore given; and if N be taken so as to have to AL the same ratio that AB^2 has to AH^2, N will be given in magnitude. Also since $AH^2 : BK^2 :: AL : LB$, and $AH^2 : AB^2 :: AL : N$, ex æquo $BK^2 : AB^2 :: LB : N$. From L draw LO, LM perpendicular to AC, CB; LO and LM are given in magnitude.

Now, because $AB^2:BK^2::AD^2:DF^2$, $N:LB::AD^2:DF^2$, so that $DF^2=\frac{LB}{N}\cdot AD^2$, and for the same reason, $DE^2=\frac{AL}{N}\cdot BD^2$. But (Loci Plani, Append. Lem. 1.) $\frac{LB}{N}\cdot AD^2+\frac{AL}{N}\cdot BD^2=\frac{LB}{N}\cdot AL^2+\frac{AL}{N}\cdot BL^2+\frac{AB}{N}\cdot DL^2$; that is, $DE^2+DF^2=LO^2+LM^2+\frac{AB}{N}\cdot DL^2$.

Join LG; then by hypothesis, LO^2+LM^2 has to LG^2 the same ratio which DF^2+DE^2 has to DG^2; and if this ratio be that of R to N, $LO^2+LM^2=\frac{R}{N}\cdot LG^2$; and therefore $DE^2+DF^2=\frac{R}{N}\cdot LG^2+\frac{AB}{N}\cdot DL^2$. But $DE^2+DF^2=\frac{R}{N}\cdot DG^2$; therefore $\frac{R}{N}\cdot LG^2+\frac{AB}{N}\cdot DL^2=\frac{R}{N}\cdot DG^2$, and $\frac{AB}{N}\cdot DL^2=\frac{R}{N}(DG^2-LG^2)$. The excess of the square of DG above the square of LG, has therefore a constant ratio to the square of DL, viz. that of AB to R. The angle DLG is therefore a right angle, and the ratio of AB to R, the ratio of equality, otherwise LD would be given in magnitude, which is contrary to the supposition. The line LG is therefore given in position; and since R is to N, that is AB to N, as the squares of LO and LM to the square of LG, therefore the square of LG, and consequently the line LG, is given in magnitude. The point G is therefore given, and also the ratio of the squares of DE and DF to the square of DG, which is the same with that of AB to N.

Hence this construction: Divide AB in L, so that AL may be to LB as the square of AH to the square of BK, and make as the square of AH to the square of AB, so AL to N; and, lastly, having drawn from L upon AC and CB the perpendiculars LO and LM, make LG perpendicular to AB, and such, that as AB to N, so the sum of the squares of LO and LM to the square of LG; G will be the point required, and the given ratio, which the squares on DF and DE have to the square on DG, will be that of AB to N.

This is the construction which follows most directly from the analysis; but it may be rendered more simple. For, since $AH^2:AB^2::AL:N$, and $BK^2:AB^2::BL:N$, therefore $AH^2+BK^2:AB^2::AB:N$. Likewise, if AG, BG be joined, $AB:N::AH^2:AG^2$, and $AB:N::BK^2:BG^2$; wherefore $AB:N::AH^2+BK^2:AG^2+BG^2$, that is, $AH^2+BK^2:AB^2::AH^2+BK^2:AG^2+BG^2$, and $AG^2+GB^2=AB^2$. The angle AGB is, therefore, a right angle, and AL:LG:LB. If, therefore, AB be divided in L, as in the preceding construction; and if LG, a mean proportional between AL and LB, be placed at right angles to AB, G will be the point required.

Cor. It is evident from the construction, that if at the points A and B we suppose weights to be placed that are as the squares of the sines of the angles CAB, CBA, L will be the centre of gravity of these weights. For AL is to LB as AC^2 to CB^2, or inversely as the squares of the sines of the angles at A and B.

25. Now, the step in this analysis by which a second introduction of the general hypothesis is avoided, is that in which the angle GLD is concluded to be a right angle. This conclusion follows from the excess of the square of DG above the square of GL, having a given ratio to the square of LD, at the same time that LD is of no determinate magnitude. For, if possible, let GLD be obtuse, (fig. 12,) and let the perpendicular from G upon AB meet AB in V, which point V is therefore given. And since the excess of the square of GD above the square of LG is equal to the square of LD, together with twice the rectangle DLV, therefore by the supposition, the square of LD, together with twice the rectangle DLV, must have a given ratio to the square of LD; the ratio of the rectangle DLV to the square of LD, that is, of VL to LD, is therefore given, so that VL being given in magnitude, LD is also given. But this is contrary to the supposition, for LD is indefinite by hypothesis; and therefore GLD cannot be obtuse, nor any other than a right angle.

The same conclusion that is here drawn immediately from the indetermination of LD, would be deduced, according to Dr Simson's method, by assuming another point D′, any how, and from the supposition, that the excess of GD'^2 above GL^2 was to LD'^2 in the same ratio that the excess of GD^2 above GL^2 is to LD^2, it would follow without much difficulty, that GLD must be a right angle, and the given ratio, a ratio of equality. The method followed above is shorter and less intricate than this last, and has, I think, the advantage of discovering more plainly the *spirit* of the analysis, and the effect which the indefinite nature of the quantities supposed indeterminate in the Porism, has in ascertaining the relation that must subsist between the magnitudes that are given, and those that are to be found.

26. This Porism may be extended to any number of lines whatsoever, and may be thus enunciated: "Let there be any number of straight lines given in position, and from any point in one of them, let perpendiculars be drawn to all the rest, a point may be found, such, that the square of the line joining it, and the point from which the perpendiculars are drawn, shall have to the sum of the squares of these perpendiculars a given ratio, which ratio is also to be found."

The analysis of the Porism, when thus generalized, is too long to be given here. * We must not,

* This Porism, in the case considered above, viz. when there are three straight lines given in position, was commu-

however, omit to take notice, that the point L, where the perpendicular from the point to be found meets the line, from which the perpendiculars are drawn to the rest, is in all cases determined by the rule suggested in the corollary, (§ 24.) For if, at the points in which the said line is intersected by the others, there be placed weights proportional to the squares of the sines of the angles of intersection, L will be the centre of gravity of these weights.

27. These Porisms facilitate the solution of the general problems from which they are derived. For, if it were proposed, three straight lines AB, AC, BC being given in position, and also a point R, (fig. 11,) to find a point D in one of the given lines AB, such, that the sum of the squares of the perpendiculars drawn from D to the other two lines, should have a given ratio to the square of DR, it is plain, that the finding of the point G in the Porism, would render the construction easy. For the squares of RD and GD, having each given

---

nicated to me several years ago, without any analysis or demonstration, by Dr Trail, Prebendary of Lisburn, in Ireland, who told me, also, that he had met with it among some of Dr Simson's papers, which had been put into his hands at the time when the posthumous works of that geometer were preparing for the press. The application of it to the second of the problems, (§ 27,) was also suggested by Dr Trail.

ratios to the sum of the squares of the perpendiculars drawn from D, have a given ratio to one another. The ratio of the lines, RD and GD themselves, is therefore given, and the points R and G being given, D is in the circumference of a circle given in position; and it is also in the straight line AB given in position; therefore it is given. The same holds, whatever be the number of lines given in position.

The same Porisms assist also in the solution of another problem. For if it were proposed to find D, so that the sum of the squares of the perpendiculars drawn from it to AC, and CB, should be equal to a given square, this would be done by finding G; and then because the sum of the squares of the perpendiculars is given, and has a given ratio to the square of DG, DG will be given, and consequently the point D. This is also true, whatever be the number of the lines.

28. The connection of the Porisms with the impossible cases of these problems, is abundantly evident; the point L being that from which, if perpendiculars be drawn to AC and CB, the sum of their squares is the least possible. For since (fig. 11.) $DF^2+DE^2:DG^2::LO^2+LM^2:LG^2$, and since LG is less than DG, LO^2+LM^2 must be less than DF^2+DE^2.

Hence also a point Q may be found, from which, if perpendiculars be drawn to the sides of the triangle ABC, the sum of the squares of these perpendiculars is less than the sum of the squares of the perpendiculars drawn to the sides of the triangle from any other point.

For if ab (fig. 13,) be any line drawn parallel to AB, and if it be divided in λ, so that $a\lambda$ may be to λb in the duplicate ratio of aC to Cb, or of AC to CB, then of all the points in the line ab, λ is that from which, if perpendiculars be drawn to the lines AC, CB, the sum of their squares is the least possible. But since $a\lambda$ is to λb as the square of AC to the square of BC, that is, as AL to LB, therefore the locus of λ is the straight line LC, joining the given points L and C. The point to be found therefore, or that from which perpendiculars being drawn to the sides of the figure, the sum of their squares is the least possible, is in the straight line LC.

For let q be any point on either side of LC, and let the line ab be drawn through q parallel to AB, meeting LC in λ, then the sum of the squares of the perpendiculars from q upon AC, CB, is greater than the sum of the squares of the perpendiculars from λ upon the same lines. Therefore adding the square of the perpendicular from q, or λ, on AB, to both, the sum of the squares of the perpendiculars from q, will be greater than the sum of the squares of the perpendiculars from λ. The point, therefore, which makes the sum of the squares of the perpendiculars drawn from it, to the sides of the triangle ABC, a minimum, is not on either side of the line LC; it is therefore in the line LC.

For the same reason, if AC be divided in L′, so that AL′ is to L′C as the square of AB to the square of BC, and if BL′ be joined, the point to be found is in BL′. It is therefore in the point Q, where the lines CL and BL′ intersect one another.

The point Q, in any other figure, may be found nearly in the same manner. Let ABCD, for instance, (fig. 14,) be a quadrilateral figure; let the opposite sides, AB and DC, be produced till they meet in E, and let *ab* be drawn parallel to AB, meeting CE in *e*, and let λ be the point in the line *ab* from which perpendiculars are drawn to the three lines BC, CD, DA, so that the sum of their squares is less, than if they were drawn from any other point, in the same line; then if weights be placed at *b*, *a* and *e*, proportional to the squares of the sines of the angles C*ba*, *ba*D, *ae*D, λ is the centre of gravity of these weights, (§ 26.) Now, these weights having given ratios to one another, the locus of the point λ, from the known properties of the centre of gravity, is a straight line Lλ, given in position. The point to be found is, therefore, in that line. For the same reason, it is in another straight line $L'\lambda'$ also given in position; and therefore it is in Q, the point of their intersection.

There are many other remarkable properties of this point, which appear sometimes in the form of Porisms, and sometimes of theorems. Of the former, some curious instances will be found in Dr Small's Demonstrations of Dr Stewart's Theorems.* Of the latter, I shall only add one, omitting the demonstration, which would lead into too long a digression. "If Q be the point in a triangle from which perpendiculars are drawn to the sides of the triangle, so that the sum of their squares is the least possible; twice the area of the triangle is a mean proportional between the sum of the squares of the sides of the triangle, and the sum of the squares of the above mentioned perpendiculars."

29. But to return to the subject of Porisms: It is evident from what has now appeared, that in some instances at least, there is a close connection between these propositions and the *maxima* or *minima*, and, of consequence, the impossible cases, of problems. The nature of this connection requires to be further investigated, and is the more interesting, that the transition from the indefinite, to the impossible cases of a problem seems to be made with wonderful rapidity. Thus, in the first pro-

* Trans. R. S. Edin. Vol. II. p. 112, &c.

position, though there be not, properly speaking, an impossible case, but only one where the point to be found goes off ad infinitum, we may remark, that if the given point F be any where out of the line HD, the problem of drawing GB equal to GF is always possible, and admits just of one solution; but if F be in the line DH, the problem admits of no solution at all, the point being then at an infinite distance, and therefore impossible to be assigned. There is, however, this exception, that if the given point be at K, in this same line DH, determined by making DK equal to DL, then every point in the line DE gives a solution, and may be taken for the point G. Here therefore the case of innumerable solutions, and the case of no solution, are as it were conterminal, and so close to one another, that if the given point be at K, the problem is indefinite, but that if it remove ever so little from K, remaining at the same time in the line DH, the problem cannot be resolved.

I had observed this remarkable affinity between cases, which in other respects are diametrically opposite, in a great variety of instances, before I perceived the reason of it, and found, that by attending to the origin which has been assigned to Porisms, I ought to have discovered it *a priori.* It is, as we have seen, a general principle, that a problem is converted into a Porism, when one, or when two, of the conditions of it, necessarily involve in them some one of the rest. Suppose, then, that two of the conditions are exactly in that state, which determines the third; then, while they remain fixed or given, should that third one be supposed to vary, or differ, ever so little, from the state required by the other two, a contradiction will ensue. Therefore if, in the hypothesis of a problem, the conditions be so related to one another as to render it indeterminate, a Porism is produced; but if, of the conditions thus related to one another, some one be supposed to vary, while the others continue the same, an absurdity follows, and the problem becomes impossible. Wherever, therefore, any problem admits both of an indeterminate and an impossible case, it is certain, that these cases are nearly related to one another, and that some of the conditions by which they are produced, are common to both. This affinity, which seems to be one of the most remarkable circumstances respecting Porisms, will be more fully illustrated, when we treat of the algebraic investigation of these propositions.

30. It is supposed above, that two of the conditions of a problem involve in them a third, and wherever that happens, the conclusion which has been deduced will invariably take place. But a Porism may sometimes be so simple, as to arise from the mere coincidence of one condition of a problem with another, though in no case whatever,

any inconsistency can take place between them. Thus, in the second of the foregoing propositions, the coincidence of the point given in the problem with another point, viz. the centre of gravity of the given triangle, renders the problem indeterminate; but as there is no relation of distance, or position, between these points, that may not exist, so the problem has no impossible case belonging to it. There are, however, comparatively but few Porisms so simple in their origin as this, or that arise from problems in which the conditions are so little complicated; for it usually happens, that a problem which can become indefinite, may also become impossible; and if so, the connection between these cases, which has been already explained, never fails to take place.

31. Another species of impossibility may frequently arise from the porismatic case of a problem, which will very much affect the application of geometry to astronomy, or any of the sciences of experiment, or observation. For when a problem is to be resolved by help of data furnished by experiment or observation, the first thing to be considered is, whether the data so obtained, be sufficient for determining the thing sought; and in this a very erroneous judgment may be formed, if we rest satisfied with a general view of the subject: For though the problem may in general be resolved from the data that we are provided with, yet these

data may be so related to one another in the case before us, that the problem will become indeterminate, and instead of one solution, will admit of an infinite number.

Suppose, for instance, that it were required to determine the position of a point F, (fig. 9,) from knowing that it was situated in the circumference of a given circle ABC, and also from knowing the ratio of its distances from two given points E and D; it is certain, that in general these data would be sufficient for determining the situation of F: But nevertheless, if E and D should be so situated, that they were in the same straight line with the centre of the given circle; and if the rectangle under their distances from that centre, were also equal to the square of the radius of the circle, then, as was shown above, (§ 12,) the position of F could not be determined.

This particular instance may not indeed occur in any of the practical applications of geometry; but there is one of the same kind which has actually occurred in astronomy: And as the history of it is not a little singular, affording besides an excellent illustration of the nature of Porisms, I hope to be excused for entering into the following detail concerning it.

32. Sir Isaac Newton having demonstrated, that the trajectory of a comet is a parabola, reduced the actual determination of the orbit of any particular

comet, to the solution of a geometrical problem, depending on the properties of the parabola, but of such considerable difficulty, that it is necessary to take the assistance of a more elementary problem, in order to find, at least nearly, the distance of the comet from the earth, at the times when it was observed. The expedient for this purpose, suggested by Newton himself, was to consider a small part of the comet's path as rectilineal, and described with an uniform motion, so that four observations of the comet being made at moderate intervals of time from one another, four straight lines would be determined, viz. the four lines joining the places of the earth and the comet, at the times of observation, across which if a straight line were drawn, so as to be cut by them into three parts, in the same ratios with the intervals of time above mentioned; the line so drawn would nearly represent the comet's path, and by its intersection with the given lines, would determine, at least nearly, the distances of the comet from the earth at the times of observation.

The geometrical problem here employed, of drawing a line to be divided by four other lines given in position, into parts having given ratios to one another, had been already resolved by Dr Wallis and Sir Christopher Wren, and to their solutions Sir Isaac Newton added three others of his own, in different parts of his works. Yet none of all these geometers observed that peculiarity in the problem which rendered it inapplicable to astronomy. This was first done by Boscovich, but not till after many trials, when, on its application to the motion of comets, it had never led to any satisfactory result. The errors it produced in some instances were so considerable, that Zanotti, seeking to determine by it the orbit of the comet of 1739, found, that his construction threw the comet on the side of the sun opposite to that on which he had actually observed it. This gave occasion to Boscovich, some years afterwards, to examine the different cases of the problem, and to remark that, in one of them, it became indeterminate; and that, by a curious coincidence, this happened in the only case which could be supposed applicable to the astronomical problem above mentioned; in other words, he found, that in the state of the data, which must there always take place, innumerable lines might be drawn, that would be all cut in the same ratio, by the four lines given in position. This he demonstrated in a dissertation published at Rome in 1749, and since that time in the third volume of his Opuscula. A demonstration of it, by the same author, is also inserted at the end of Castillon's Commentary on the Arithmetica Universalis, where it is deduced from a construction of the general problem, given by Thomas

Simpson, at the end of his Elements of Geometry.* The proposition, in Boscovich's words, is this: "Problema quo quæritur recta linea quæ quatuor rectas positione datas ita secet, ut tria ejus segmenta sint invicem in ratione data, evadit aliquando indeterminatum, ita ut per quodvis punctum cujusvis ex iis quatuor rectis duci possit recta linea, quæ ei conditioni faciat satis." †

It is needless, I believe, to remark, that the proposition thus enunciated is a Porism, and that it was discovered by Boscovich, in the same way, in which I have supposed Porisms to have been first discovered by the geometers of antiquity. I shall add here a new analysis of it, conducted according to the method of the preceding examples, and to which the following lemma is subservient.

Lemma I. Fig. 15.

32. If two straight lines, AE and BF, be cut by three other straight lines, AB, CD, and EF, given in position, and not all parallel to one another, into segments having the same given ratio, they will intercept between them segments

* Elements, p. 243, Edit. 3. Simpson's solution is remarkably elegant, but no mention is made in it, of the indeterminate case.

† Jos. Boscovich Opera, Bassano. Tom. III. p. 331.

of the lines given in position, viz. AB, CD, EF, which will also have given ratios to one another.*

---

* DEMONSTRATION.—Through C and E draw CH and EG, both parallel to AB, and let them meet BG, parallel to AE, in H and in G. Let GF and HD be joined; and because AC is to CE, that is, BH to HG as BD to DF, by hypothesis, DH is parallel to GF, and has also a given ratio to it, viz. the ratio of GB to BH, or of EA to AC. Take GK equal to HD, and join EK, and the triangle EGK will be equal to the triangle CHD, and therefore the angle KEG is given, and likewise the angle KEF; and since the ratio of GK to KF is given, if from K there be drawn KL parallel to EG, meeting EF in L, the ratio of EL to LF will be given. But the ratio of EL to LK is given, because the triangle EKL is given in species; therefore the ratio of FL to LK is given; and the angle FLK being also given, the triangle FKL is given in species, as also the triangle FGE. The angle FGE being therefore given, the triangle KGE is given in species, and EG has therefore given ratios to EK and EF. But EG is equal to AB, and EK to CD, therefore AB, CD, and EF have given ratios to one another Q. E. D.

Hence to find the ratios of AB, CD, and EF; in EF take any part EL, and make as AC is to CE, so EL to LF; through L draw LK parallel to EG or AB, meeting EK, drawn through E parallel to CD in K; then if FK be drawn meeting EG in G, the ratios required are the same with the ratios of the lines EG, EK, EF. This is evident from the preceding investigation.

If it be required to find the position of the line AE, drawn through the point A, so as to be cut by CD and EF in a given ratio; draw A*c*, any how, cutting DC in *c*, and pro-

Prop. VI. Porism. Fig. 16.

33. Three straight lines being given in position, a fourth line, also given in position, may be found, such, that through any point whatever a straight line may be drawn, which will intersect these four lines, and will be divided by them into three segments, having given ratios to one another.

Let AB, CD, EF be the three lines given in position, and OL the line to be found, and $\alpha\delta$ a given line, of which the segments $\alpha\beta$, $\beta\gamma$, $\gamma\delta$ have given ratios to one another.

Let A be a given point in the line AB, and

---

duce A*c* to *e*, so that A*c* may be to *ce* in the ratio which AC is to have to CE; let *e*E be drawn parallel to DC, intersecting FE in E, and if AE be joined, it is the line required.

Hence the converse of the lemma is easily demonstrated, viz. that if AE and BF be two lines that are cut proportionally by the three lines AB, CD, EF; and if AB and EF, the parts of any two of these last, intercepted between AE and BF, be also cut proportionally, any how, in *b* and *f*, and if *bf* be joined, meeting the third line in *d*, *bf* will be cut in the same proportion with AE or BF. For, if not, let *bf'* be drawn from *b*, meeting CD in *d'*, and EF in *f'*, so that *bd'* : *d'f'* : : AC : CE ; then, by the lemma, A*b* : AB : : E*f* : EF; and by supposition, A*b* : AB : : E*f'* : EF, therefore E*f'* = E*f*, which is impossible. Therefore, &c.

suppose, that AO is drawn from it, intersecting the lines CD, EF, and OL in the points C, E, and O, and divided at these points into the segments AC, CE, EO, having the same ratios to one another, with the given segments $\alpha\beta$, $\beta\gamma$, $\gamma\delta$ of the line $\alpha\delta$. Then, because the lines CD, EF are given in position, and also the point A, the line AE is given in position and magnitude, (§ 32,) and therefore also EO, which has a given ratio to AE; the point O is therefore given.

Again, let B be any point whatever in AB, and let BL be drawn, according to the hypothesis of the Porism, so as to be divided in the points D, F, and L, where it intersects the lines CD, EF, and OL into the parts, BD, DF, and FL, having the same ratios with the parts $\alpha\beta$, $\beta\gamma$, $\gamma\delta$.

Let also BG be drawn equal and parallel to AE, and let EG be joined; EG will therefore be parallel to AB, and will be given in position; and if GF be drawn, it will make given angles with EG and EF, because, by the preceding *lemma*, the ratio of AB to EF, that is, of EG to EF is given. Through L draw LN parallel to BG, meeting GF produced in N.

Then because the triangles BFG, LFN are similar, GF is to FN as BF to FL, that is, in a given ratio; and therefore, since FG also makes given angles with the two straight lines EG and EF, given in position, the point N is in a straight line, given in position, and passing through E, viz. EN.

Now, since BF is to FL as BG to LN, and also as AE or BG to EO, LN and EO are equal, and being also parallel, OL is parallel to EN, that is, to a line given in position; and the point O, in OL, is given, therefore OL is given in position; which was to be found.

Construction. From any two given points, A and B′, in the line AB, draw AE and B′F′ intersecting CD and EF in C, E, D′ and F′, so that AC may be to CE, and B′D′ to D′F′ in the same given ratio of $\alpha\beta$ to $\beta\gamma$, (§ 32.) Produce also AE to O, and B′F′ to L′, so that AE may be to EO, and B′F′ to F′L′ in the same given ratio of $\alpha\gamma$ to $\gamma\delta$. If OL′ be joined, it will be the line required.

For let B′ be any point whatsoever in AB, and as AB′ to AB, so let OL′ be to OL, and let BL be drawn, cutting CD′ and EF′ in D and F, the line BL is divided in these points, similarly to the given line $\alpha\delta$. For since the two lines AO and B′L′ are divided similarly by the three lines AB′, CD′ and OL′, and since two of these last, AB′ and OL′, are also divided similarly to one another by the three lines AO, B′L′ and BL, BL will be divided in D, in the same ratio wherein B′L′ is divided in D′, or AO in C, (Lem. 1. Conv.). In the same way, BL is divided in F, in the same ratio wherein AO is divided in E; BL is therefore similarly divided to AO, or to $\alpha\delta$, which was to be demonstrated.

34. Hence it is plain, "If two similarly divided lines, as AO and BL, be drawn any how, and if straight lines AB, CD, EF, OL, be drawn through the points of division of these lines, innumerable lines may be placed between the lines AB, CD, EF and OL, which will be divided by them, similarly to the lines AO, and BL." For, by what is here demonstrated, every line which cuts any two of the lines AB, CD, &c. proportionally, will also cut the others proportionally, and will be cut by them into segments having the same ratio to one another, with the segments of the lines AO and BL.

From this it follows, that the astronomical problem above mentioned, becomes a Porism, and is indeterminate, in the case when the observations of the comet are not very distant from one another. For on this supposition, the arches described by the earth, and by the comet during the time in which the observations are made, will not differ much from two straight lines; and these lines will be divided similarly to one another, because each of them will be divided into parts, proportional to the intervals of time between the observations. The places of the earth, at the times of the observations, may therefore be nearly represented by the points A, C, E and O, in the straight line AO, and those of the comet by the points B, D, F and L, in the straight line BL, these lines AO and BL being divided both into parts having the same ratios. The position of BL therefore is not given, since, by the Porism, it may be any line whatever, which cuts the two lines, AB and OL, in a certain ratio.

It is also to be remarked, that in order to render this, or any other geometrical problem, of no use in questions where the data are furnished by observation, and are consequently liable to some inaccuracy, it is not necessary, that the problem should be reduced exactly to the porismatic case; for even on a near approach to that case, a very small error in the data will produce so great an error in the conclusion, that no dependence can be had upon its accuracy.

This will be made evident, in the present instance, by considering how the construction of the Porism is subservient to the solution of the other cases of the problem. Suppose that four lines, AB, CD, EF, RS, (fig. 16,) are given in position, and that it is required to draw a straight line that shall be divided by these lines into parts having the ratios of the given lines $\alpha\beta$, $\beta\gamma$, $\gamma\delta$. Let KT be that line, and assuming the points A and B′, and drawing the lines AO, B′L′, so that they may be similarly divided to the line $\alpha\delta$, as in the construction of the Porism, then if OL be joined, it will be given in position, and the extremity K, of the line KT, will be in the line OL′, by the Porism ; but it is also in the line RS ; it is therefore given. Now, by the lemma, AT is to TB′ as OK to KL′, and the lines OK and KL′ being given, the ratio of AT to TB′ is given, so that T is given, and therefore TK is given in position Q. E. I.

Now, it is evident, that if RS make a small angle with OL, any error in the determination of that angle will make a great variation in the position of the point K. A small change in it may, for instance, make RS parallel to OL, and consequently will throw off K to an infinite distance, so that the line, which is sought, will be impossible to be found ; and, in general, the variation of the position of K, corresponding to a given variation in the angle RKO, will be, *cæteris paribus*, inversely as the square of the sine of that angle. The nearer, therefore, that the problem is to the Porism, the less is the solution of it to be depended on, and the more does it partake of the indefinite character of the latter.

35. Sir Isaac Newton has extended the hypothesis of the problem from which the preceding Porism is derived, and has formed from it one more general, which he has also resolved, with a view to its application in astronomy. It is this : " To describe a quadrilateral given in species, that shall

have its angles upon four straight lines given in position." *

As it is evident, that the former problem is but a particular case of this last, it is natural to expect, that a Porism is also to be derived from it, or that the lines given in position may be such, that the problem will become indeterminate. On attempting the analysis, I have accordingly found this conjecture verified; the investigation depending on a lemma similar to that which is prefixed to the preceding proposition.

Lemma II. Fig. 17.

If two triangles ABC, DEF, similar to a given triangle, be placed with their angles on three straight lines given in position, so that the equal angles in both the triangles may be upon the same straight lines, the ratios of the segments of these straight lines, intercepted between the two triangles, that is, of AD, BE, and CF, are given.†

---

* Prin. Math. Lib. I. lem. 27.

† Demonstration.—Complete the parallelogram under AC and AD, viz. AG, and on DG describe the triangle DGH, similar and equal to the triangle ABC. Join FG, BH and HE. Through G also, draw GK, equal and parallel to HE, and join CK; CK will be equal and parallel to BE, and the triangle CGK equal to the triangle BHE.

PROP. VII. PORISM. FIG. 18.

36. Three straight lines being given in position, a fourth may be found, which will also be given in

---

The angle GCK is therefore given, being equal to the given angle HBE; and the angle GCF being given, the angle FCK is also given.

The triangles DHE, DGF are similar; for the angles FDE, GDH being equal, the angles FDG, EDH are likewise equal; and also, by supposition, FD being to DE as GD to DH, FD is to DG as DE to DH. The angle FGD is therefore equal to the angle EHD, and FG is also to EH, or to KG, as FD to DE, or as GD to DH.

But if GL be drawn parallel to HD, the angle KGL will be equal to the angle EHD, that is, to the angle FGD, and therefore the angle KGF to the angle LGD or GDH; and it has been shown, that FG is to GK as GD to DH; therefore the triangle FGK is similar to the triangle GDH, and is given in species.

Draw GM perpendicular to CF, and GN making the angle MGN equal to the angle FGK or GDH, and let GM be to GN in the given ratio of FG to GK, or of GD to DH. Join CN and NK. Then, because MG:GN::FG:GK, MG:FG::GN:GK; and the angle MGF being equal NGK, the triangles MGF, NGK are similar, and therefore GNK is a right angle. But since the ratio of MG to GN is given, and also of MG to GC, the triangle CGM being given in species, the ratio of GC to GN is given, and CGN being also a given angle, because each of the angles CGM, MGN is given, the triangle CGN is given in species, and consequently the ratio of CG to CN is given. The angle NCK is therefore given; and the angle CNK is likewise

position, and will be such, that innumerable quadrilaterals, similar to the same given quadrilateral, may be described, having their angles placed, in the same order, on the four straight lines given in position.

Let AD, BE, CF, be the three straight lines given in position, and *ablc* a given quadrilateral. Let A be a given point in the line AD, and let ABLC be a quadrilateral, similar to the given quadrilateral *ablc*, placed, so that the angles of the triangle ABC, similar to the given triangle *abc*, may be, one of them, at the given point A, and the other two, on the lines BE and CF. The points B and C, and the triangle ABC, will therefore be

---

given, each of the angles CNG, GNK being given, therefore the triangle CNK is also given in species. The ratio of CN to CK is therefore given, and since the ratio of CN to CG is also given, the ratio of CG to CK is given, and the triangle CGK given in species. The angle KGC is therefore given, and the angle KGF being also given, the angle CGF is given, and consequently the ratio of CG to CF. The ratios of the lines CG, CK and CF to one another, that is, of AD, BE and CF to one another, are therefore given. Q. E. D.

Cor. Hence also it appears, how a triangle given in species may be described, having its angles on three straight lines given in position, and one of the angles at a given point in one of the lines. The solution of this problem is therefore taken for granted, in the analysis of the Porism, though, for the sake of brevity, the construction is omitted.

given, (Lemma 2. Cor.) and consequently the triangle CBL will also be given in position and magnitude, and the point L will be given. The line to be found must pass through L; let it be LM; let M be any point in it whatsoever, and let MEDF be a quadrilateral similar to the given quadrilateral *ablc*, having its angles on the four lines LM, CF, BE and AD, the angle at M being equal to the angle CLB, &c.

Complete the parallelogram AG, under CA, AD, and on DG describe the quadrilateral GDHN, similar and equal to the quadrilateral ABLC; join BH and LN, and it is evident, that the three lines CG, BH, and LN are all equal, and parallel to AD, and are all given in position. Join also AL, DN, DM, MN, and FG.

Because the two quadrilaterals DEMF, DHNG are similar, the angle FDM is equal to the angle GDN, and therefore the angle GDF to the angle NDM. For the same reason also, GD : DF : : ND : DM, and therefore the triangles GDF, NDM are similar, and the angle FGD equal to the angle MND, and FG : MN : : GD : DN, so that FG has a given ratio to MN.

But because the triangles ABC, DEF are similar, CG has a given ratio to CF, (Lem. 2.) so that the angle GCF being given, the triangle CGF is given in species, and FG has to GC a given ratio; now, FG was also shown to have to MN a given ratio; therefore MN has a given ratio to CG, that is, to LN.

Again, since the triangle CGF is given in species, the angle CGF is given, and CGD being also a given angle, the angle FGD is given, and therefore MND, which is equal to it. But the angle LND is given, therefore the angle LNM is given; and it was shown, that MN has a given ratio to NL, therefore the angle MLN is given; now, the point L, and the line LN, are given in position; therefore LM is also given in position, which was to be found.

The construction for finding LM is obvious. Take A and D, two given points in one of the lines given in position, and place the two triangles ABC, DEF similar to the given triangle *abc*, so that two of their equal angles may be at A and D, and the other equal angles on the lines BE and CF, (Lem. 2. Cor.). On BC and EF, describe the triangles BLC, FEM, similar to the triangle *cbl*; if LM be drawn, it will be the line required.

From the analysis it also follows, that the quadrilaterals described with their angles, on the four straight lines given in position, as supposed in the Porism, will intercept between them segments of these lines, having given ratios to one another.

37. This Porism may also be extended to figures of any number of sides, and may be enunciated more generally thus: "A rectilineal figure of any

number of sides, as *m*, being given, and three straight lines being also given in position, $m-3$ straight lines may be found given in position, so that innumerable rectilineal figures may be described, similar to the given rectilineal figure, and having their angles on the straight lines given in position."

Hence also this theorem: "If any two rectilineal figures be described similar to one another, and if straight lines be drawn, joining the equal angles of the two figures, innumerable rectilineal figures may be described, which will have their angles on these lines, and will be similar to the given rectilineal figures; and the segments of the lines given in position, intercepted between any two of these figures, will have constantly the same ratio to one another."

As a *Locus*, the same proposition admits of a very simple enunciation, and has a remarkable affinity to that with which Euclid appears to have introduced his first book of Porisms. "If three of the angles of a rectilineal figure, given in species, be upon three straight lines given in position, the remaining angles of the figure will also be on straight lines, given in position."

If the rectilineal figures here referred to be such, as may be inscribed in a circle, or in similar curves of any kind, agreeably to the hypothesis of the problem,* by which these last Porisms were suggested, we shall have a number of other Porisms respecting straight lines given in position, which cut off, from innumerable such curves, segments that are given in species. A great field of geometrical investigation is, therefore, opened by the two preceding propositions, which, however, we must at present be content to have pointed out.

38. A question nearly connected with the origin of Porisms still remains to be resolved, namely, from what cause has it arisen, that propositions which are in themselves so important, and that actually occupied so considerable a place in the ancient geometry, have been so little remarked in the modern? It cannot indeed be said, that propositions of this kind were wholly unknown to the moderns before the restoration of what Euclid had written concerning them; for beside Boscovich's proposition, of which so much has been already said, the theorem which asserts, that in every system of points there is a centre of gravity, has been shown above to be a Porism; and we shall see hereafter, that many of the theorems in the higher geometry belong to the same class of propositions. We may add, that some of the elementary propositions of geometry want only the proper form of enunciation to be perfect Porisms. It is not therefore strictly

* Prin. Math. Lib. 1. Prop. 29.

true, that none of the propositions called Porisms have been known to the moderns; but it is certain, that they have not met from them, with the attention they met with, from the ancients, and that they have not been distinguished as a separate class of propositions. The cause of this difference is undoubtedly to be sought for in a comparison of the methods employed for the solution of geometrical problems in ancient, and in modern times.

In the solution of such problems, the geometers of antiquity proceeded with the utmost caution, and were careful to remark every particular case, that is to say, every change in the construction, which any change in the state of the data could produce. The different conditions from which the solutions were derived, were supposed to vary one by one, while the others remained the same; and all their possible combinations being thus enumerated, a separate solution was given, wherever any considerable change was observed to have taken place.

This was so much the case, that the Sectio Rationis, a geometrical problem of no great difficulty, and one of which the solution would be dispatched, according to the methods of the modern geometry, in a single page, was made, by Apollonius, the subject of a treatise consisting of two books. The first book has seven general divisions, and twenty-four cases; the second, fourteen general divisions, and seventy-three cases, each of which cases is separately considered. Nothing, it is evident, that was any way connected with the problem, could escape a geometer, who proceeded with such minuteness of investigation.

The same scrupulous exactness may be remarked in all the other mathematical researches of the ancients; and the reason doubtless is, that the geometers of those ages, however expert they were in the use of their analysis, had not sufficient experience in its powers, to trust to the more general applications of it. That principle which we call the *law of continuity*, and which connects the whole system of mathematical truths by a chain of insensible gradations, was scarcely known to them, and has been unfolded to us, only by a more extensive knowledge of the mathematical sciences, and by that most perfect mode of expressing the relations of quantity, which forms the language of algebra; and it is this principle alone which has taught us, that though in the solution of a problem, it may be impossible to conduct the investigation without assuming the data in a *particular* state, yet the result may be perfectly *general*, and will accommodate itself to every case with such wonderful versatility, as is scarcely credible to the most experienced mathematician, and such as often forces him to stop, in the midst of his calculus, and to look back, with a mixture of diffidence and admiration, on the unforeseen harmony of his conclusions. All this was unknown to the ancients; and therefore they had no resource, but to apply their analysis separately to each particular case, with that extreme caution which has just been described; and in doing so, they were likely to remark many peculiarities, which more extensive views, and more expeditious methods of investigation, might perhaps have induced them to overlook.

39. To rest satisfied, indeed, with too general results, and not to descend sufficiently into particular details, may be considered as a vice that naturally arises out of the excellence of the modern analysis. The effect which this has had, in concealing from us the class of propositions we are now considering, cannot be better illustrated than by the example of the Porism discovered by Boscovich, in the manner related above. Though the problem from which that Porism is derived, was resolved by several mathematicians of the first eminence, among whom also was Sir Isaac Newton, yet the Porism which, as it happens, is the most important case of it, was not observed by any of them. This is the more remarkable, that Sir Isaac Newton takes notice of the two most simple cases, in which the problem obviously admits of innumerable solutions, viz. when the lines given in position are either all parallel, or all meeting in a point, and these two hypotheses he therefore expressly excepts. Yet he did not remark, that there are other circumstances

which may render the solution of the problem indeterminate, as well as these; so that the porismatic case considered above, escaped his observation: And if it escaped the observation of one who was accustomed to penetrate so far into matters infinitely more obscure, it was because he satisfied himself with a general construction, without pursuing it into its particular cases. Had the solution been conducted after the manner of Euclid or Apollonius, the Porism in question must infallibly have been discovered.

But I have already extended this paper to too great a length; so that, leaving the use of algebra in the investigation of Porisms, to be treated of on another occasion, I shall conclude with a remark from Pappus, the truth of which, I would willingly flatter myself, that the foregoing observations have had some tendency to evince: "Habent autem Porismata subtilem et naturalem contemplationem, necessariam et maxime universalem, atque iis, qui singula perspicere et investigare valent, admodum jucundam."

FINIS.

OBSERVATIONS

ON THE

TRIGONOMETRICAL TABLES

OF THE

BRAHMINS.

ON THE

TRIGONOMETRY

OF THE

BRAHMINS.*

---

1. IN the second volume of the Asiatic Researches, an extract is given from the Surya Siddhanta, the ancient book which has been long, though obscurely, pointed out as the source of the astronomical knowledge of the Brahmins. The Surya Siddhanta is in the Sanscrit language: It is one of the Sastras, or inspired writings of the Hindoos, and is called the Jyotish, or Astronomical, Sastra. It professes, as we learn from Mr Davis, the ingenious translator, to be a revelation from heaven, communicated to Meya, a man of great sanctity, about four millions of years ago, toward the close of the Satya Jug, or of the Golden Age of the Indian mythologists; a period at which man is said to have been incomparably better than he is at present; when

---

* From the Transactions of the Royal Society of Edinburgh, Vol. IV. (1798.)—ED.

his stature exceeded twenty-one cubits, and his life extended to ten thousand years.

Interwoven, however, with all these extravagant fictions, this singular book contains a very sober and rational system of astronomical calculation; and even the principles and rules of trigonometry, a science of all others the most remote from fable, and the least susceptible of poetical decoration. It is on the construction of the tables contained in this trigonometry, that I now beg leave to offer a few remarks.

2. It is necessary to begin with observing, that the circumference of the circle is here divided into 360 equal parts, each of which is again subdivided into 60, and so on. The same division was followed by the Greek mathematicians; and this coincidence is the more to be remarked, that it relates to a matter of arbitrary arrangement, and one by no means necessarily connected with the properties of the circle. There are indeed some very obvious properties of that curve, that make it, though not necessary, at least convenient, that the number of parts, into which the circumference is divided, should be a number divisible both by 3 and by 4, that is, that it should be a multiple of 12; but nothing more precise can be determined from the nature of the curve itself. The agreement of two nations, therefore, in dividing the circumference of the circle precisely in the same manner, as it can-

not well be attributed to chance, must be supposed to result from some communication having taken place between them, if it were not that another very probable cause may be assigned for it. In Greece, and no doubt in every other country, the division of the circle, into equal parts, is of a much older date than the origin of trigonometry, and must be as ancient as the first circular instruments used for measuring angles in the heavens. The inventors of those instruments naturally sought to make the divisions on them correspond to the space which the sun described daily in the ecliptic; and they could easily discover, without any very precise knowledge of the length of the solar year, that this might be nearly effected by making each of them the 360th part of the whole circumference. Accordingly the famous circle of Osymandias, in Egypt, described by Herodotus, was divided into 360 equal parts.

This principle may therefore have guided the astronomers, both of the East and of the West, to the same division of the circle, without any intercourse having taken place between them. It has certainly directed the Chinese in their division, though it has led them to adopt one different from the Hindoo and Egyptian astronomers. They divide the circle into $365\frac{1}{4}$ parts, which can have no other origin than the sun's annual motion: and some such division as this, may perhaps have been the first that was employed by other nations, who changed it however to the number 360, which nearly answered the same purpose, and had besides the great advantage of being divisible into many aliquot parts. The Chinese, again, with whom the sciences became stationary almost from their birth, have never attempted to improve on the method that first occurred to them.

3. The next thing to be mentioned, is also a matter of arbitrary arrangement, but one in which the Brahmins follow a method peculiar to themselves. They express the radius of the circle in parts of the circumference, and suppose it equal to 3438 minutes, or 60ths of a degree. In this they are quite singular. Ptolemy, and the Greek mathematicians, after dividing the circumference, as we have already described, supposed the radius to be divided into 60 equal parts, without seeking to ascertain, in this division, any thing of the relation of the diameter to the circumference: and thus, throughout the whole of their tables, the chords are expressed in sexagesimals of the radius, and the arches in sexagesimals of the circumference. They had therefore two measures, and two units; one for the circumference, and another for the diameter. The Hindoo mathematicians, again, have but one measure and one unit for both, viz. a minute of a degree, or one of those parts whereof the circumference contains 21600. From this identity of

measures, they derive no inconsiderable advantage in many calculations, though it must be confessed, that the measuring of a straight line, the radius, or diameter of a circle, by parts of a curve line, namely, the circumference, is a refinement not at all obvious, and has probably been suggested to them by some very particular view, which they have taken, of the nature and properties of the circle. As to the accuracy of the measure here assigned to the radius, viz. 3438 of the parts of which the circumference contains 21600, it is as great as can be attained, without taking in smaller divisions than minutes, or 60ths of a degree. It is true to the nearest minute, and this is all the exactness aimed at in these trigonometrical tables. It must not however be supposed, that the author of them meant to assert, that the circumference is to the radius, either accurately or even very nearly, as 21600 to 3438. I have shown, in another place,* from the Institutes of Akbar, that the Brahmins knew the ratio of the diameter to the circumference to great exactness, and supposed it to be that of 1 to 3.1416, which is much nearer than the preceding. Calculating, as we may suppose, by this or some other proportion, not less exact, the authors of the tables found, that the radius contained in truth 3437′. 44″. 48‴, &c.; and as the fraction of a minute is

---

* Page 163.

here more than a half, they took, as their constant custom is, the integer next above, and called the radius 3438 minutes. The method by which they came to such an accurate knowledge of the ratio of the diameter to the circumference, may have been founded on the same theorems which were subservient to the construction of their trigonometrical tables.*

4. These tables are two, the one of sines, and the other of versed sines. The sine of an arch they call *cramajya* or *jyapinda*, and the versed sine *utcramajya*. They also make use of the cosine or *bhujajya*. These terms seem all to be derived from the word *jya*, which signifies the chord of an arch, from which the name of the radius, or sine of 90°, viz. *trijya*, is also taken. This regularity in their trigonometrical language, is a circumstance not unworthy of remark. But what is of more consequence to be observed, is, that the use of sines, as it was unknown to the Greeks, who calculated by help of the chords, forms a striking difference between the Indian trigonometry and theirs. The use of the sine, instead of the chord, is an improvement which our modern trigonometry owes, as we have hitherto been taught to believe, to the Arabs; and it is certainly one of the acquisitions which the mathematical sciences made, when, on

* See Note § 6.

their expulsion from Europe, they took refuge in the East. But whether the Arabs are the authors of this invention, or whether they themselves received it, as they did the numerical characters, from India, is a question, which a more perfect knowledge of Hindoo literature will probably enable us to resolve.

No mention is made in this trigonometry, of tangents or secants; a circumstance not wonderful, when we consider that the use of these was introduced in Europe no longer ago than the middle of the sixteenth century. It is, on the other hand, not a little singular, that we should find a table of versed sines in the Surya Siddhanta; for neither the Greek nor the Arabian mathematicians, had any such, nor had we, in modern Europe, till after the time of Petiscus, who wrote about the end of the century just mentioned.

5. Next, as to the extent and accuracy of these tables. The first of them exhibits the sines to every twenty-fourth part of the quadrant, that is, the sine of 3° 45′, and of all the multiples of that arch, viz. 7° 30′, 11° 15′, &c. up to 90°. The table of versed sines does the same. In each, the sine, or versed sine, is expressed in minutes of the circumference, but without any fractions of a minute, either decimal or sexagesimal; and, agreeably to the observation already made, when the fraction that ought to have been set down is greater than

$\frac{1}{2}$, the integer next greater is placed in the table. Thus the sine 3° 45′ being, when accurately expressed in their way, 224′ 49″, is put down 225′; and so of the rest. The numbers, therefore, in these tables, are only so far exact as never to differ more than half a minute from the truth, and this very limited degree of accuracy gives no doubt to their trigonometry the appearance of an infant science: But when, on the other hand, we consider the principles and rules of their calculations, rather than the numbers actually calculated, we find the marks of a science in full vigour and maturity: and we will acknowledge, that the Hindoo mathematicians did not satisfy themselves with the degree of accuracy above mentioned, from any incapacity of attaining to greater exactness.

Their rules for constructing their tables of sines, may be reduced to two, viz. the one for finding the sine of the least arch in the table, that of 3° 45′, and the other for finding the sines of the multiples of that arch, its triple, quadruple, &c. Both of these Mr Davis has translated, judging very rightly, that it was impossible to give two more curious specimens of the geometrical knowledge of the Hindoo philosophers; the first is extracted from a commentary on the Surya Siddhanta; the other from the Surya Siddhanta itself.

6. With respect to the first, the method proceeds by the continual bisection of the arch of 30°

and correspondent extractions of the square root, to find the sine and cosine of the half, the fourth part, the eighth part, and so on, of that arch. The rule, when the sine of an arch is given, to find that of half the arch, is precisely the same with our own: "The sine of an arch being given, find the cosine, and thence the versed sine, of the same arch: then multiply half the radius into the versed sine, and the square root of the product is the sine of half the given arch." Now, as the sine of 30°, was well known to those mathematicians to be half the radius, it was of consequence given: thence by the rule just laid down, was found the sine of 15°, then of 7° 30′, and lastly of 3° 45′, which is the sine required. Thus the sine of 3° 45′ would be found equal to 224′ 44″, as above observed, and, the sine of 7° 30′, equal to 448′ 39″, and, taking the nearest integers, the first was made equal to 225′, and the second to 449′.*

* By such continual bisections, the Hindoo mathematicians, like those of Europe before the invention of infinite series, may have approximated to the ratio of the diameter to the circumference, and found it to be nearly that of 1 to 3.1416 as above observed. A much less degree of geometrical knowledge than they possessed, would inform them, that small arches are nearly equal to their sines, and that the smaller they are, the nearer is this equality to the truth. If, therefore, they assumed the radius equal to 1, or any number at pleasure, after carrying the bisection of the arch of

7. When, by the bisections that have just been described, the sine of 3° 45′ or of 225′, was found equal to 225′, the rest of the table was constructed by a rule, that, for its simplicity and elegance, as well as for some other reasons, is entitled to particular attention. It is as follows: "Divide the first jyapinda, 225′ by 225; the quotient 1, deducted from the dividend, leaves 224′, which added to the first jyapinda, or sine, gives the second, or the sine of 7° 30′, equal to 449′. Divide the second jyapinda, which is thus found, by 225, and deduct 2, the nearest integer to the quotient, from the former remainder 224′, and this new remainder 222′, added to the second jyapinda, will give the third jyapinda equal to 671′. Divide this last by 225, and subtract 3, the nearest integer to the quotient, from the former remainder 222′, and there will be left 219′, which, added to the third

---

30°, two steps farther than in the above construction, they would find the sine of the 384th part of the circle, which, therefore, multiplied by 384, would nearly be equal to the circumference itself, and would actually give the proportion of 1 to 3.14159, as somewhat greater than that of the diameter to the circumference. By carrying the bisections farther, they might verify this calculation, or estimate the degree of its exactness, and might assume the ratio of 1 to 3.1416 as more simple than that just mentioned, and sufficiently near to the truth.

jyapinda, gives the fourth; and so on unto the twenty-fourth or last."

It is not immediately obvious on what geometrical principle this rule is founded, but a slight change in the enunciation will remove the difficulty. The remainder, it must be observed, from which the quotient is always directed to be taken away, is the difference between the two sines last computed; and hence the rule may be expressed more generally: Divide any sine by 225, and subtract the quotient, or the integer nearest the quotient, from the difference between that sine and the sine next less; the remainder is the difference between the same sine and the sine next greater; and therefore if it be added to the former, will give the latter. If then, (fig. 19,) GA, GC, GE, be three contiguous arches in the table, of which the differences AC, CE, of consequence are equal, and of which the sines are AB, CD, and EF, the rule, as last stated, gives us $CD - AB - \frac{CD}{225}$, for the difference between CD and EF, and therefore $EF = CD + CD - AB - \frac{CD}{225} = 2CD - \frac{CD}{225} - AB$, and also $EF + AB = CD\left(2 - \frac{1}{225}\right) = CD\left(\frac{449}{225}\right)$. But 225 is the sine of the arch 3° 45′, and 449 of twice that arch, as already shown; and, therefore, according to this rule, if there be three arches of which the common difference is 3° 45′, the sine of the mean arch will always have to the sum of the sines of the extreme arches. a given ratio, that namely, which the sine of 3° 45′ has to the sine of twice 3° 45′, or of 7° 30′; now, this is a true proposition; and therefore we are in possession of the principle on which the Hindoo canon is constructed.

8. The geometrical theorem, which is thus shown to be the foundation of the trigonometry of Hindostan, may also be more generally enunciated. "If there be three arches in arithmetical progression, the sine of the middle arch is to the sum of the sines of the two extreme arches, as the sine of the difference of the arches to the sine of twice that difference." This theorem is well known in Europe; it is justly reckoned a very remarkable property of the circle; and it serves to show, that the numbers in a table of sines constitute a series, in which every term is formed exactly in the same way, from the two preceding terms, viz. by multiplying the last by a certain, constant number, and subtracting the last but one from the product.

9. Now, it is worth remarking, that this property of the table of sines, which has been so long known in the East, was not observed by the mathematicians of Europe till about two hundred years ago. The theorem, indeed, concerning the circle, from which it is deduced, under cne shape or an-

other, has been known to them from an early period, and may be traced up to the writings of Euclid, where a proposition nearly related to it forms the 97th of the Data: "If a straight line be drawn within a circle given in magnitude, cutting off a segment containing a given angle, and if the angle in the segment be bisected by a straight line produced till it meet the circumference; the straight lines, which contain the given angle, shall both of them together have a given ratio to the straight line which bisects the angle." This is not precisely the same with the theorem which has been shown to be the foundation of the Hindoo rule, but differs from it only by affirming a certain relation to hold among the chords of arches, which the other affirms to hold of their sines. It is given by Euclid as useful for the construction of geometrical problems; and trigonometry being then unknown, he probably did not think of any other application of it. But what may seem extraordinary is, that when, about 400 years afterwards, Ptolemy, the astronomer, constructed a set of trigonometrical tables, he never considered Euclid's theorem, though he was probably not ignorant of it, as having any connection with the matter he had in hand. He, therefore, founded his calculations on another proposition, containing a property of quadrilateral figures inscribed in a circle, which he seems to have investigated on purpose, and which is still distinguished by his name. This proposition comprehends in fact Euclid's, and of course the Hindoo theorem, as a particular case ; and though this case would have been the most useful to Ptolemy, of all others, it appears to have escaped his observation ; on which account he did not perceive that every number in his tables might be calculated from the two preceding numbers, by an operation extremely simple, and every where the same ; and therefore his method of constructing them is infinitely more operose and complicated than it needed to have been.

Not only did this escape Ptolemy, but it remained unnoticed by the mathematicians, both Europeans and Arabians, who came after him, though they applied the force of their minds to nothing more than to trigonometry, and actually enriched that science by a great number of valuable discoveries. They continued to construct their tables by the same methods which Ptolemy had employed, till about the end of the sixteenth century, when the theorem in question, or that on which the Hindoo rule is founded, was discovered by Vieta. We are, however, ignorant by what train of reasoning that excellent geometer discovered it ; for, though it is published in his Treatise on Angular Sections, it appears there not with his own demonstration, but with one given by an ingenious mathematician of our own country, Alexander Anderson of Aberdeen. It was then regarded as a theorem entirely new, and I know not that any of the geometers of that age remarked its affinity to the propositions of Euclid and Ptolemy. It was soon after applied in Europe, as it had been so many ages before in Hindostan, and quickly gave to the construction of the trigonometrical canon all the simplicity which it seems capable of attaining. From all this, I think it might fairly be concluded, even if we had no knowledge of the antiquity of the Surya Siddhanta, that the trigonometry contained in it is not borrowed from Greece or Arabia, as its fundamental rule was unknown to the geometers of both those countries, and is greatly preferable to that which they employed.

10. Considerable light may perhaps hereafter be thrown on this argument, if it be found that the Surya Siddhanta contains a demonstration of this rule. It does not appear, however, from the fragment we are in possession of, that any explanation of the rule is given, either in that work or in the commentary. Indeed, I am not certain that the Surya Siddhanta contains any thing but rules and maxims, or that the author of it condescends to give any demonstrations of the propositions which he enunciates. He may have felt himself relieved from the necessity of doing so, by his claim to inspiration ; and as he probably valued himself more on the character of a prophet, than of a geometer,

he may rather have inclined to exercise the faith, than the reason, of his disciples.

However that be, by the rule above explained, the Brahmins have computed a set of tables, limited, indeed, in their accuracy, but extremely simple and compendious. The rule is easily remembered by one who has been accustomed to numerical calculation, and is such, that, by help of it, he may at any time compose for himself a complete set of trigonometrical tables, in a few hours, without the assistance of any book whatever. For the purpose of rendering it thus simple, the contrivance of measuring the radius, and all the sines, in parts of the circumference, seems to have been adopted: if we follow any other method, the rule, though it remain the same in reality, will assume a form much less easy to be retained in the memory. * It

* This seems to me the most probable reason that can be assigned for the measuring of the radius, and the other straight lines in the circle, in parts of the circumference. It is remarkable that the Hindoos should have been thus led, at so early a period, to put in practice a method, the same in the most material point, with one which has been but lately suggested in Europe, as an important improvement in trigonometrical calculation. In the Phil. Trans. for 1783, Dr Hutton of Woolwich proposed to divide the circumference, not into degrees, as is usually done, but into decimals of the radius; and he has pointed out how the present trigonometrical tables might be accommodated to this new division, with the least possible labour, in a paper which dis-

has the appearance, like many other things in the science of those eastern nations, of being drawn up by one who was more deeply versed in the subject than may be at first imagined, and who knew much more than he thought it necessary to communicate. It is probably a compendium, formed by some ancient adept in geometry, for the use of others who were merely practical calculators.

11. If we were not already acquainted with the high antiquity of the astronomy of Hindostan, nothing could appear more singular than to find a system of trigonometry, so perfect in its principles, in a book so ancient as the Surya Siddhanta. The antiquity of that book, the oldest of the Sastras, can scarce be accounted less than 2000 years before our era, even if we follow the very moderate system of Indian chronology laid down by Sir William Jones. * Now, if we suppose its antiquity to be no

---

plays that intimate acquaintance with the resources, both of the numerical and algebraic calculus, for which he is so much distinguished. His plan is, in one respect, the same with the Hindoo method, for it uses the same unit to express both the circumference and the diameter; in another respect it differs from it, viz. in making the radius the unit, while the other assumes for an unit the 360th part of the circumference. Dr Hutton's plan has never been executed, though it certainly would be of advantage to have, besides the ordinary trigonometrical tables, others constructed according to that plan.

* Asiatic Researches, Vol. II. p. 111, &c.

higher than this, though it bears in itself internal marks of an age still more remote, * yet it will sufficiently excite our wonder to find it contain the principles of a science, of which the first rudiments are not older in Greece than 130 years before our era. The bare existence of trigonometrical tables, though they belong undoubtedly to a very elementary branch of science, yet argues a state of greater advancement in the mathematics than may at first be imagined, and necessarily supposes the applica-

---

* The obliquity of the ecliptic is stated at 24° in the Surya Siddhanta, as in all the other astronomical tables of the Hindoos which we are yet acquainted with. (P. 133.) Mr Davis concludes from this, (Asiatic Researches, Vol. II. p. 238,) that if the obliquity diminish, at the rate of 50″ in a hundred years, the Surya Siddhanta is at present about 3840 years old, which goes back nearly 2000 years before the Christian era. But the diminution of the obliquity of the ecliptic, is supposed considerably too rapid in this calculation. According to Mayer it is 46″ in a century; and according to Lagrange, (Mém. Berlin, 1782,) at a medium no more than 30″. This last is most to be depended on, as it proceeds on an accurate inquiry into the law of the secular variation of the obliquity, that variation being by no means uniform. Let us, however, take the mean, viz. 38″, and the obliquity at the beginning of the present century having been 23°, 28′, 41″, we shall have 5000 years for the age of the Surya Siddhanta, reckoned from that date, or about 3300 years before Christ, which is near the era of the Calyougham.

tion of geometrical reasoning to some of the more difficult problems of astronomy and geography.

As long as the surveying of land, and the ordinary mensuration of surfaces and solids, are the only practical arts to which the geometer applies his speculations, he will naturally content himself with constructing his figures and plans by means of a scale, and an instrument for measuring angles, as by doing so he may attain to all the accuracy he can desire. But when, in the figures that are to be thus delineated, the sides happen to be extremely unequal, and some of the angles very acute, or very obtuse, graphical operations become inaccurate, and a very small error in the measuring of one thing produces an enormous error in the estimation of some other. Lines, therefore, that extend over a great tract of the earth's surface, and much more those that extend to the heavens, cannot be compared with the smaller lines, which we have an opportunity of measuring, by the bare construction of triangles and parallelograms; and when ever such comparisons are to be made, some other method must be sought for. It was precisely in such circumstances, that the inventive genius of Hipparchus suggested the application of arithmetic to ascertain those ratios among the sides and angles of figures, which pure geometry afforded no method of expressing. This union of geometry and arithmetic did not happen, however, till each of these sciences separately had made great progress; for before the days of Hipparchus, Euclid, Archimedes, and Apollonius, had all flourished in succession, and had produced those immortal works, of which the lustre has not been obscured by the highest improvements of later ages. In the progress of science, therefore, the invention of trigonometry is to be considered as a step of great importance, and of considerable difficulty. It is an application of arithmetic to geometry, with which we are now too familiar, to perceive all the merit of the inventor; but a little reflection will convince us, that he, who first formed the idea of exhibiting, in arithmetical tables, the ratios of the sides and angles of all possible triangles, and contrived the means of constructing such tables, must have been a man of profound thought, and of extensive knowledge. However ancient, therefore, any book may be, in which we meet with a system of trigonometry, we may be assured, that it was not written in the infancy of science.

12. As we cannot, therefore, suppose the art of trigonometrical calculation to have been introduced till after a long preparation of other acquisitions, both geometrical and astronomical, we must reckon far back from the date of the Surya Siddhanta, before we come to the origin of the mathematical sciences in India. In Greece, the constellations were first represented on the sphere, if we take a

medium between the chronology of Newton, and that which is now generally received, about 1140 years before the Christian era;* and Hipparchus invented trigonometry 130 years before the same era. Even among the Greeks, therefore, an interval, of at least 1000 years, elapsed from the first observations in astronomy, to the invention of trigonometry; and we have surely no reason to suppose, that the progress of knowledge has been more rapid in other countries.

A thousand years therefore must be added to the age of the Surya Siddhanta, which we suppose here to be 2000 before Christ, in order that we may reach the origin of the sciences in Hindostan, and this brings us very nearly to the celebrated era of the Calyougham, to which M. Bailly has already referred the construction of the astronomical tables of that country. And here, I cannot help observing, in justice to an author, of whose talents and genius the world has been so unseasonably and so cruelly deprived, that his opinions, with respect to this era, appear to have been often misunderstood.

---

* The sphere of Chiron and Musæus was constructed, according to Newton, about the year 936 before Christ. (Newton's Chron. Chap. i. § 30.) According to the system generally received, the ancient sphere, described by Eudoxus, was constructed about 1350 years before Christ. (Dr Playfair's Chronology, p. 37.) The medium is 1143.

It certainly was not his intention to assert, that the Calyougham was a *real* era, considered with respect to the mythology of India, or even that at so remote a period the religion of Brahma had an existence. The religious and civil institutions of Hindostan, as they now exist, may be all posterior to this date, and their antiquity is probably to be determined from principles that are not the objects of astronomical discussion. All, I think, therefore, that Bailly meant to affirm, and certainly all that is necessary to his system, is, that the Calyougham, or the year 3102 before our era, marks a point in the duration of the world, before which the foundations of astronomy were laid in the East, and those observations made, from which the tables of the Brahmins have been composed.

On this, however, and on many more of the particulars of the history of those remote ages, great additional light will undoubtedly be thrown, by the complete translation of the Surya Siddhanta. From the specimen which Mr Davis has given, we can neither doubt of the importance of such a work, nor of his abilities to execute it; and we trust, that, to the zeal and liberality of our brethren of the Asiatic Society, the learned world will soon be indebted for the possession of this inestimable treasure.

FINIS.

THEOREMS

RELATING TO

THE FIGURE OF THE EARTH.

ON THE

FIGURE OF THE EARTH.*

1. THE observations which have been made to determine the magnitude and figure of the earth, have not hitherto led to results completely satisfactory. They have indeed demonstrated the compression or oblateness of the terrestrial spheroid, but they have left an uncertainty as to the quantity of that compression, extending from about the one hundred and seventieth, to the three hundred and thirtieth part of the radius of the equator. Between these two quantities, the former of which is nearly double of the latter, most of the results are placed, but in such a manner that those best entitled to credit are much nearer to the least extreme than to the greatest. Sir Isaac Newton, as is well known, supposing the earth to be of uniform density, assigned for the compression at the poles $\frac{1}{230}$, nearly a mean between the two limits just

* From the Transactions of the Royal Society of Edinburgh, Vol. V. (1805.)—ED.

mentioned; and it is probable, that, if the compression is less than this, it is owing to the increase of the density toward the centre. Boscovich, taking a mean from all the measures of degrees, so as to make the positive and negative errors equal, found the difference of the axes of the meridian $=\frac{1}{248}$. By comparing the degrees measured by Father Leisganig in Germany, with eight others that have been measured in different latitudes, Lalande finds $\frac{1}{311}$, and, suppressing the degree in Lapland, which appears to err in excess, $\frac{1}{331}$ for the compression. Laplace makes it $\frac{1}{321}$; Sejour $\frac{1}{307}$, and, lastly, Carouge and Lalande $\frac{1}{300}$.

These results, which reduce the eccentricity of the meridians so much lower than was once supposed, agree well with the observations of the length of the pendulum made in different latitudes. Were the earth a homogeneous body, Sir Isaac Newton demonstrated, that the diminution of gravity under the equator would be $=\frac{1}{230}$, expressed by the same fraction with the compression at the poles. Clairaut made afterwards a very important addition to this theorem; for he showed, that, if the earth be not homogeneous, but have a density that varies with any function of the distance from the centre, the two fractions, expressing the compression at the poles, and the diminution of gravity at the equator, when added together, must be of the same amount as in the homogeneous spheroid, that is, must be $=\frac{2}{230}$ or $\frac{1}{115}$. Now, the seconds pendulum is concluded, from the best and most recent observations, to be longer at the pole than at the equator by $\frac{1}{185}$, and this, taken from $\frac{1}{115}$, leaves $\frac{1}{304}$ for the compression at the poles.

2. But though $\frac{1}{300}$, or some fraction not very different from it, should be admitted as the most probable value of the compression, or ellipticity, as it is called, of the terrestrial spheroid, it still remains to be explained, why all the observations, considering the care with which they have been made, do not agree more nearly with this conclusion. Among the causes that may be assigned for this inconsistency, though unavoidable mistakes, and the imperfection of instruments, must come in for a part, there can be little doubt that local irregularities in the direction of gravity have had the greatest share in producing it. Of these irregularities, that which arises from the attraction of mountains has had its existence proved, and its quantity, in one case, ascertained, by the very accurate observations of the present Astronomer-Royal

at Scheballien in Perthshire. We may trace the operation of this cause in many of the degrees that have been actually measured. Thus, in the degree at Turin, when divided into two parts, and each estimated separately, that which was to the north of the city, and pointed toward Monte Rosa, the second of the Alps in elevation, and the first perhaps in magnitude, was found greater in proportion than that toward the south, the plummet having been attracted by the mountain above mentioned, and the zenith made of consequence to recede toward the south. There are no doubt situations in which the measurement of a small arch might, from a similar cause, give the radius of curvature of the meridian infinite, or even negative.

But there is another kind of local irregularity in the direction of gravity, that may also have had a great effect in disturbing the accuracy of the measurement of degrees. The irregularity I mean is one arising from the unequal density of the materials under and not far from the surface of the earth; and this cause of error is formidable, not only because it may go to a great extent, but because there is not any visible mark by which its existence can always be distinguished. The difference between the primary and secondary strata is probably one of the chief circumstances on which this inequality depends. The primary strata, especially if we include among them the granite, may often have three times the specific gravity of water, whereas the secondary, such as the marly and argillaceous, frequently have not more than twice the specific gravity of that fluid. Suppose, then, that a degree is measured in a country where the strata are all secondary, and happens to terminate near the junction of these with the primitive or denser strata, the line of which junction we shall also suppose to lie nearly east and west; the superior attraction of the denser strata must draw the plummet toward them, and make the zenith retire in the opposite direction; thus diminishing the amplitude of the celestial arch, and increasing, of consequence, the geodetical measure assigned to a degree. From suppositions, no way improbable, concerning the density and extent of such masses of strata, I have found that the errors, thus produced, may easily amount to ten or twelve seconds.

3. While we continue to draw our conclusions, about the figure of the earth, from the measurement of single degrees, there appears to be no way of avoiding, or even of diminishing, the effects of these errors. But if the arches measured are large, and consist each of several degrees, though there should be the same error in determining their celestial amplitudes, the effect of that error, with respect to the magnitude and figure of the earth, will become inconsiderable, being spread out over a greater interval; and it is, therefore, by the com-

parison of two such arches that the most accurate result is likely to be obtained. But, in pursuing this method, since the arches measured cannot be treated as small quantities, or mere fluxions of the earth's circumference, the calculation must be made by rules quite different from those that have been hitherto employed. These new rules are deduced from the following analysis.

4. Let the ellipsis ADBE (fig. 20) represent a meridian passing through the poles D and E, and cutting the equator in A and B. Let C be the centre of the earth, AC, the radius of the equator, $=a$, and DC, half the polar axis, $=b$. Let FG be any very small arch of the meridian, having its centre of curvature in H; join HF, HG cutting AC in K and L. Let φ be the measure of the latitude of F, or the measure of the angle AKF, expressed, not in degrees and minutes, but in decimals of the radius 1; then the excess of the angle ALG above AKF, that is, the angle LHK or GHF will be $=\dot{\varphi}$, and therefore $FG=\dot{\varphi}\times FH$. Also, if the elliptic arch $AF=z$, $FG=\dot{z}=\dot{\varphi}\times FH$.

But FH, or the radius of curvature at F, is =

$$\frac{a^2b^2}{(a^2-a^2\sin^2\varphi+b^2\sin^2\varphi)^{\frac{3}{2}}}=a^2b^2(a^2-a^2\sin^2\varphi+b^2\sin^2\varphi)^{-\frac{3}{2}},\text{ as}$$

is demonstrated in the conic sections. Therefore, if c be the compression at the poles, or the excess of a above b, $b^2=a^2-2ac+c^2$, or because c is small in comparison of a, if we reject its powers higher than the first, $b^2=a^2-2ac$, and $\text{FH}=a^3(a-2c)$ $(a^2-a^2\sin^2\varphi+a^2\sin^2\varphi-2ac\sin^2\varphi)^{-\frac{3}{2}}=a^3(a-2c)(a^2-2ac\sin^2\varphi)^{-\frac{3}{2}}$.

But $(a^2-2ac\sin^2\varphi)^{-\frac{3}{2}}=a^{-3}(1-\frac{2c}{a}\sin^2\varphi)^{-\frac{3}{2}}=a^{-3}$ $(1+\frac{3c}{a}\sin^2\varphi)$ nearly, rejecting, as before, the terms that involve c^2, &c. Hence $\text{FH}=(a-2c)(1+\frac{3c}{a}\sin^2\varphi)$ $=a-2c+3c\sin^2\varphi$.

Now $\dot{z}=\dot{\varphi}\times\text{FH}$, therefore $\dot{z}=\dot{\varphi}(a-2c+3c\sin^2\varphi)$ $=(a-2c)\dot{\varphi}+3c\dot{\varphi}\sin^2\varphi$. But $\sin^2\varphi=\frac{1-\cos 2\varphi}{2}$, therefore $\dot{z}=(a-2c)\dot{\varphi}+\frac{3}{2}c\dot{\varphi}-\frac{3c\dot{\varphi}}{2}\cos 2\varphi$, and taking the fluent $z=\left(a-\frac{c}{2}\right)\varphi-\frac{3c}{4}\sin 2\varphi$. To this value of z no constant quantity is to be added, because it vanishes when $z=o$.

Therefore an arch of the meridian, extending from the equator to any latitude φ, is $=a\varphi-\frac{c}{2}\left(\varphi+\frac{3}{2}\sin 2\varphi.\right)$.

5. This theorem is also easily applied to measure an arch of the meridian, intercepted between any two parallels of the equator.

Thus, if MN be any arch of the meridian, φ' the latitude of M, one of its extremities, and φ'' that of N, its other extremity, we have

$AM = a\varphi' - \frac{c}{2}\left(\varphi' + \frac{3}{2}\sin 2\varphi'\right)$, and

$AN = a\varphi'' - \frac{c}{2}\left(\varphi'' + \frac{3}{2}\sin 2\varphi''\right)$. Therefore the arch

$MN = a(\varphi'' - \varphi') - \frac{c}{2}\left((\varphi'' - \varphi') + \frac{3}{2}\sin 2\varphi'' - \frac{3}{2}\sin 2\varphi'\right)$.

6. If, therefore, MN be an arch of several degrees of the meridian, the length of which is known by actual measurement, and also the latitude of its two extremities M and N, this last formula gives us an equation, in which a and c are the only unknown quantities. In the same manner, by the measurement of another arch of the meridian, an equation will be found, in which a and c are likewise the only unknown quantities. By a comparison, therefore, of these two equations, the values of a and c, that is of the radius of the equator, and its excess above the polar axis, may be determined.

Thus, if l be the length of an arch measured, m the coefficient of a, and n of c, computed by the last formula; and if l' be the length of any other arch, m' the coefficient of a, and n' of c, computed in the same manner, we have $ma - nc = l$,

and $m'a - n'c = l'$.

Whence $a = \frac{n'l - nl'}{mn' - m'n}$; $c = \frac{m'l - ml'}{mn' - m'n}$ and $\frac{c}{a} = \frac{m'l - ml'}{n'l - nl'}$.

It may be useful, in the numerical calculation, to observe also that $c = \frac{ma - l}{n}$.

7. The arch of the meridian, which was mea-

sured in Peru, compared with that measured in France, will afford an example of the application of these formulas.

The amplitude of the arch measured in Peru was 3°. 7′. 1″, and its length 176940 toises. To reduce this to the level of the sea, above which it was elevated 1226 toises, 66 toises must be subtracted, and again 12 toises added to adapt it to the mean temperature of the atmosphere. Thus corrected it is 176886 toises. The arch measured begun 36″ north of the equator, and extended to the parallel of 3°. 6′. 25 south; we shall suppose it to have begun under the equator, and to have extended to the parallel of 3°. 7′. 1″, a supposition which can produce no sensible error, and will somewhat simplify the calculation. Thus φ, in the preceding formula, is an arch of 3°. 7′. 1″ expressed in decimals of the radius 1, and so we have $m = .0544009$, $n = .1086408$, and $l = 176886$.

Again, the amplitude of the whole arch measured in France from Dunkirk to Perpignan is 8°. 20′. 2″½, and its length 475496 toises. The northern extremity of this arch is in latitude 51°. 2′. 1″, and the southern in 42°. 41′. 58″½ Hence $\varphi'' = .8907045$, and $\varphi' = .7452459$, and therefore $m' = .1454586$, $n' = .0585735$, $l = 475496$.

Therefore, $a = \dfrac{n'l - nl'}{mn' - m'n} = 3273325$ toises;

$c = \dfrac{m'l - ml'}{mn' - m'n} = \quad 10917$ toises,

$$\text{and } \frac{c}{a} = \frac{1}{300} \text{ nearly.}$$

Wherefore also the longer axis of the meridian is to its conjugate, or a is to b as 300 to 299.

This proportion agrees well with that which was already pointed out as the most probable result, from the comparison of single degrees, and from observations of the pendulum. As these conclusions are obtained by different methods, they tend greatly to confirm one another.

8. From this, too, it seems highly probable, that the uncertainty which yet remains with respect to the true figure of the earth will be entirely removed by the measurement of some other considerable arches of the meridian. Such an arch will be furnished by the survey of Great Britain begun by General Roy, and still continued in a style of accuracy so much superior to any other system of geometrical operations that has ever yet been executed. In drawing the conclusions from observations made with such exactness, it may be necessary to employ a more accurate approximation than has been done in the preceding *formulæ*, by retaining the second power of c. The equations to be resolved will thus become of the second order, but as the unknown quantities can be nearly found by the solution of a simple equation, the farther approximation to their true values will be accompanied with no difficulty.

9. Concerning this farther approximation, it may be useful, however, to remark, that if c^2 be retained, its coefficient in the formula of § 4, will be $\frac{1}{16a}\left(\varphi+\frac{15}{4}\sin 4\varphi\right)$; and therefore, in the formula of § 5, it will be $\frac{1}{16a}\left(\varphi''-\varphi'+\frac{15}{4}(\sin 4\varphi''-\sin 4\varphi')\right)$.

If then the quantity $\frac{1}{16a}\left(\varphi''-\varphi'+\frac{15}{4}(\sin 4\varphi''-\sin 4\varphi')\right)$ computed for any arch of the meridian, be put $=g$, and the same, computed for any other arch, be $=g'$, the equations of § 6 will become

$$ma-nc+\frac{gc^2}{a}=l, \text{ and}$$

$$m'a-n'c+\frac{g'c^2}{a}=l'.$$

10. Here if we put d for the value of a, as given by the formula $\frac{n'l-nl'}{mn'-m'n}$; and h for the value of c, as given by the formula $\frac{m'l-ml'}{mn'-m'n}$, also v for the correction to be made on d, and u for the correction to be made on h, so that $a=d+v$, and $c=h+u$, by substituting these values of a and c in the two last equations, we have $mv-nu+\frac{g(h+u)^2}{d+v}=0$, and

$$m'v-n'u+\frac{g'(h+u)^2}{d+v}=0.$$

Hence, rejecting all the terms that involve v^2, u^2, or uv, we have $dmv-dnu+gh^2+2ghv=0$,

and $dm'v-dn'u+g'h^2+2g'hv=0$.

Therefore, $v=\frac{(ng'-n'g)h^2}{(n'm-nm')d+(gn'-g'n)2h}$, also

$$u=\frac{g'h^2(dm+2gh)-gh^2(dm'+2g'h)}{dn'(dm+2gh)-dn(dm'+2g'h)}.$$

And, again, by rejecting those terms that are small in comparison of the rest, $v=\frac{h^2(ng'-n'g)}{d(n'm-nm')}$, and

$$u=\frac{h^2(g'm-gm')}{d(n'm-nm')}.$$

Thus v and u are found, and of consequence $d+v$ and $h+u$, that is a and c, without neglecting any terms that are not of an order less than $\frac{c^2}{a}$; and when it is considered that $\frac{c^2}{a}$ is less than $\frac{1}{22500}$, it will readily be allowed that it is quite unnecessary to carry the approximation farther.

11. The same thing that renders the comparison of large arches of the meridian useful for lessening the effect of errors arising from irregularities in the direction of gravity, makes it serve to diminish the effect of all the errors of the astronomical observations at the extremities of the arches, from whatever cause they arise. They are all diffused over a greater interval, and have an effect proportionally less in diminishing the accuracy of the last conclusion.

12. The measurement therefore of large arches of the meridian, especially if performed in distant countries, is likely to furnish the best *data* for as-

certaining the true figure of the earth; and on this account extensive and accurate surveys, such as that above mentioned, are no less interesting to science, in general, than conducive to national utility. The survey of this Island, when completed, will furnish an arch of the meridian, beginning at the same parallel where that measured in France terminates, and nearly of the same extent, so that the length of an arch of more than 16°, or almost a twentieth of the earth's circumference, will become known. The different portions of this arch compared with one another, or with the arch measured in Peru, will afford a variety of *data* for determining the true figure of the earth.

But surveys of the kind now referred to, afford likewise other materials from which the solution of this great geographical problem may be deduced. These are chiefly of two sorts, viz. the magnitude of arches, either of the curves perpendicular to the meridian, or of the circles parallel to the equator. Examples of the first of these have been given by General Roy and Mr Dalby; the observations which follow are directed toward both.

13. With respect to the measurement of arches perpendicular to the meridian, it may be observed, that the directions of gravity at different points of such arches do not intersect one another at all, unless the distances of those points from the said meridian be very small. On this account the measure-

ment of a large arch perpendicular to the meridian would involve in it considerable difficulty ; to avoid which it is necessary that the arch measured be but small, or one that does not greatly exceed a single degree. Such measurements are of course obnoxious to all the errors that arise from the deflection of the plumb-line, and cannot therefore furnish *data* for determining the figure of the earth, equally valuable with those which may be derived from large arches of the meridian. The method of determining the figure of the earth, from degrees of the perpendicular to the meridian, is not however without its advantages, and in certain circumstances is preferable to any other that proceeds by the measurement of arches equally small. This method is twofold ; as a degree of the meridian may be compared with a degree of the perpendicular to it in the same latitude ; or two degrees perpendicular to the meridian, in different latitudes, may be compared with one another. The advantages peculiar to each will appear from the following investigation.

14. Let it be required to find the axes of an elliptic spheroid, from comparing a degree of the meridian in any latitude with a degree of the curve perpendicular to the meridian in the same latitude.

Let the ellipsis ADBE (fig. 20.) represent a meridian, of which a degree is measured at F. Let the perpendicular to the meridian in F meet the less axis DE in R. Then R will be the centre of curvature of the circle cutting the meridian at right angles in F; for at any point in that circle indefinitely near to F, the direction of the plumb-line, or of gravity, as it always passes through the axis DE, will cut DE in R; it will therefore also intersect FR in R, so that R is the centre, and RF the radius, of curvature of the perpendicular to the meridian. Let H be the centre of curvature of the meridian itself at F: draw FO perpendicular to DE, and let the latitude of F, or the angle $OFR=\varphi$. Also let $AC=a$, $CD=b$, and $a-b=c$, as before.

Then from the nature of the ellipsis, $FO=\frac{a^2\cos\varphi}{\sqrt{a^2\cos^2\varphi+b^2\sin\varphi^2}}$, and because $\sin FRO:1::FO:FR$, that is, $\cos\varphi:1::FO:FR$, $FR=\frac{a^2}{\sqrt{a^2\cos^2\varphi+b^2\sin\varphi^2}}$; and this, therefore, is the radius of curvature of the section of the spheroid perpendicular to the meridian at F. But the radius of curvature of the meridian at F, that is $FH=\frac{a^2b^2}{\sqrt{a^2\cos^2\varphi+b^2\sin^2\varphi}}$, therefore $FR:FH::\frac{a^2}{(a^2\cos\varphi^2+b^2\sin\varphi^2)^{\frac{1}{2}}}:\frac{a^2b^2}{(a^2\cos\varphi^2+b^2\sin\varphi^2)^{\frac{3}{2}}}$, and dividing both by $\frac{a^2}{(a^2\cos\varphi^2+b^2\sin\varphi^2)^{\frac{1}{2}}}$, we have $FR:FH::a^2\cos\varphi^2+b^2\sin\varphi^2:b^2$.

15. If then D be the length of a degree of the meridian at F, and D′ the length of a degree of the

circle at right angles to it, $D' : D :: a^2 \cos\varphi^2 + b^2 \sin\varphi^2 : b^2$ and $\frac{D'}{D} = \frac{a^2\cos\varphi^2 + b^2\sin\varphi^2}{b^2} = \frac{a^2}{b^2}\cos\varphi^2 + \sin\varphi^2$. Hence $\frac{D'}{D} - \sin\varphi^2 = \frac{a^2}{b^2}\cos\varphi^2$ and $\frac{a}{b} = \frac{\sqrt{\frac{D'}{D} - \sin\varphi^2}}{\cos\varphi}$.

This last formula, therefore, gives the ratio of a to b when D,D′ and φ are known.

16. To find a and b themselves, if $m=57.2957$, &c. or the number of degrees in the radius, so that $mD' = FR = \frac{a^2}{(a^2\cos\varphi^2 + b^2\sin\varphi^2)^{\frac{1}{2}}}$, and since it has been already shown that $\frac{a^2}{b^2} = \frac{\frac{D'}{D} - \sin\varphi^2}{\cos\varphi^2}$, or $b^2 = \frac{a^2\cos\varphi^2}{\frac{D'}{D} - \sin\varphi^2}$,

therefore $mD' = \frac{a^2}{\left(a^2\cos\varphi^2 + \frac{a^2\cos\varphi^2}{\frac{D'}{D} - \sin\varphi^2} \times \sin\varphi^2\right)^{\frac{1}{2}}} =$

$\frac{a}{\cos\varphi\left(1 + \frac{\sin\varphi^2}{\frac{D'}{D} - \sin\varphi^2}\right)^{\frac{1}{2}}}$ and $a = mD'\cos\varphi\sqrt{1 + \frac{\sin\varphi^2}{\frac{D'}{D} - \sin\varphi^2}}$.

Now, $1 + \frac{\sin\varphi^2}{\frac{D'}{D} - \sin\varphi^2} = \frac{\frac{D'}{D}}{\frac{D'}{D} - \sin\varphi^2} = \frac{1}{1 - \frac{D}{D'}\sin\varphi^2}$,

therefore $a = \frac{mD'\cos\varphi}{\sqrt{1 - \frac{D}{D'}\sin\varphi^2}}$.

17. This value of a is very convenient for logarithmical calculation; for if $\sin\varphi\sqrt{\frac{D}{D'}}$ be computed,

it will always be less than 1, because D′ is greater than D, and therefore may be taken for the sine of an arch ψ, of which arch $\sqrt{1-\frac{D}{D'}\sin\varphi^2}$ will of course be the cosine, so that $a=\frac{mD'\cos\varphi}{\cos\psi}$.

The same method may be used for finding $\frac{a}{b}$ from the formula in § 15.

In the same manner that a has been found, we will obtain $b=\frac{mD'\cos\varphi^2}{\left(1-\frac{D}{D'}\sin\varphi^2\right)\sqrt{\frac{D'}{D}}}$.

If we examine these formulas in the extreme cases, viz. when $\varphi=90°$, and when $\varphi=0$, we shall have in the former case $a=\frac{0}{0}$, because $\cos\varphi=0$, and also $D'=D$, so that $1-\frac{D}{D'}\sin^2\varphi=0$. Here therefore a is indefinite, and may be of any magnitude whatever; and it is evident that this is the result which the formula ought to give: because at the pole, or when $\varphi=90°$, the perpendicular to the meridian is itself a meridian, and therefore the measurement of the two degrees, D and D′, is but the same with the measurement of one degree.

When $\varphi=0$, that is at the equator, the circle perpendicular to the meridian is the equator itself, and we have then $a=mD'$, a being determined in this case by the degree of the equator alone. Here also we have $\frac{a}{b}=\sqrt{\frac{D'}{D}}$, which is known to be true.

18. The preceding formulas may be rendered more simple, if we aim only at an approximation, which indeed is all that is necessary in this inquiry. Since c denotes the compression, or since $a-c=b$, and therefore $a^2-2ac=b^2$ nearly, consequently the radius of curvature of the meridian at F, that is

$$mD=\frac{a^2(a^2-2ac)}{(a^2-2ac\sin\varphi^2)^{\frac{3}{2}}}=\frac{a^3(a-2c)}{a^3(1-\frac{2c}{a}\sin\varphi^2)^{\frac{3}{2}}}=(a-2c)(1-$$

$\frac{3c}{a}\sin\varphi^2)$, or $mD=a-2c+3c\sin\varphi^2$. In the same manner $mD'=a+c\sin\varphi^2$. From these equations we obtain, rejecting always the higher powers of c,

$$c=\frac{m(D'-D)}{2\cos\varphi^2},\ a=mD'-\frac{m(D'-D)\sin\varphi^2}{2\cos\varphi^2};\text{ and }\frac{c}{a}=\frac{D'-D}{2D'\cos\varphi^2}.$$

These formulas may be transformed into others a little more convenient for computation, by putting $\sec\varphi^2$ instead of $\frac{1}{\cos\varphi^2}$, and $\tan\varphi^2$ instead of $\frac{\sin\varphi^2}{\cos\varphi^2}$; we have then,

$$c=\frac{m}{2}(D'-D)\sec\varphi^2,$$

$$a=mD'-\frac{m}{2}(D'-D)\tan\varphi^2, \text{ and}$$

$$\frac{c}{a}=\frac{(D'-D)}{2D'}\sec\varphi^2.$$

19. We may apply these formulas to the computation of $\frac{c}{a}$, &c. from the degrees of the meridian and perpendicular, measured in the south of England. We find, in one example, (Phil. Trans.

1795, p. 537,) that D=60851 fathoms, D′=61182, the latitude, or φ being $=50°.41′$. From this $\frac{c}{a}=\frac{D'-D}{2D'\cos^2\varphi}=\frac{331}{2\times 61182(\cos 50°.41')^2}=\frac{1}{148.4}$, which is nearly the same result with that deduced in the passage just referred to. Indeed, the solution of this problem, contained in the Trigonometrical Survey, is quite unexceptionable; and the theorems here offered are not given as containing a more accurate solution, but one that is in some respects more simple.

The above compression, if the remarks already made be well founded, is much too great, being more than double of what was obtained from comparing the whole arch of the meridian measured in France with the whole of that measured in Peru. At the same time it is right to observe, that all the other comparisons of the degrees of the meridian, with those of the curve perpendicular to it, made from the observations in the south of England, agree nearly in giving the same oblateness to the terrestrial spheroid. For this circumstance, it is certainly not easy to account; the unparalleled accuracy with which the whole of the measurement has been conducted, makes it in the highest degree improbable that it arises from any error; and even if errors were to be admitted, it is not likely that they should all fall on the same side. The authors of the Trigonometrical Survey seem willing, therefore, to give up the elliptic figure of the earth, (*Ibid.* p. 527;) but before we abandon that very natural and simple hypothesis, it may perhaps be worth while to attend to the following considerations.

20. In the part of England where the measures we are now treating of have been taken, the strata are of chalk, and though of great extent, are bordered, on all sides that we have access to examine by strata much denser and more compact. Toward the west, the chalk is succeeded by limestone, and that limestone by the primitive schistus and granite of the west of Devonshire and of Cornwall. On the east we may suppose that something of the same kind takes place, though the sea prevents us from observing it, as the chalky and argillaceous beds extend in this direction to the coast, and probably to some distance beyond it. Now the meridian of Greenwich may be considered as dividing the tract of country occupied by these lighter strata, into two parts, in such a manner, that the plummet being carried to a distance from it, either east or west, approaches to the denser strata, and is of course attracted by them, so that the zenith is forced back, as it were, to the meridian of Greenwich, and does not recede from it, in the heavens, at so great a rate as the plummet itself does on the earth. Hence the longitudes from this meridian, estimated by the arches in the heavens, intercepted

between the zenith and the said meridian, will appear less than they ought to do; and too much space on the surface of the earth will, of consequence, be assigned as the measure of a degree. In this way D' is made too great; and we may suppose the circumstances such that D, on going north or south, is not enlarged in the same proportion; hence $\frac{D'-D}{D'}$ will be augmented, and of course $\frac{c}{a}$ will be represented as too great. This explanation may perhaps appear very hypothetical, and it is certainly proposed merely as a hypothesis. It is a hypothesis, too, that lays claim only to a temporary indulgence, as it is proposed at the very moment when it may be brought to the trial, and when, by a further continuation of the survey toward the north, it will probably be determined how far the distribution of the strata of this country affects the direction of gravity. It will, indeed, be curious to remark what irregularities take place on advancing into the denser strata of the north. The limestone and sandstone strata of the middle part of the island will succeed to the chalk of the south, the primitive and denser strata still occupying the west, at least at intervals, as in Wales, Cumberland, and Galloway. Further to the north, that is, beyond the Tay, the strata become entirely primitive, most of them of the densest kind, and in the interior of the island, with a very few exceptions, continue the same to its most northern extremity. In the survey of Britain, therefore, several situations must occur where the plummet, passing from lighter to denser strata, ought to give indications of some irregularities in the direction of the gravitating force. It will be seen hereafter how far these conjectures are verified by experience.

21. A remark, that is in no danger of being reckoned hypothetical, is, that the conclusion derived from the comparison of degrees of the meridian, with degrees of the circle perpendicular to it, becomes of necessity more liable to error as we advance into higher latitudes. The reason is, that whatever error is committed in determining the magnitude of D′—D, must be multiplied into the square of the secant of the latitude, in order to give its full effect in changing the value of the fraction $\frac{c}{a}$. For it has been shown, that $\frac{c}{a}=\frac{1}{2}\left(\frac{D'-D}{D'}\right)\sec^2\varphi$; now, if we suppose the error committed in ascertaining D′—D to be in all cases the same, the error of the fraction $\frac{D'-D}{D'}$ will also be in all cases nearly the same, the denominator D′ being but little affected either by the supposed error, or by the change of latitude. But this error, which may thus be considered as a constant quantity, when multiplied into $\frac{1}{2}\sec^2\varphi$, gives the variation or

error in $\frac{c}{a}$, which error therefore increases, *cæteris paribus*, as the square of the secant of the latitude, so that, on approaching the pole, it increases without limit, and is ultimately infinite. Comparisons of this kind may therefore be expected to give results the more accurate the nearer they are to the equator, under which circle they will be the most accurate of all. Here, again, however, another circumstance must be taken into consideration, viz. that the method of ascertaining the differences of longitude by the convergency of the meridians, so convenient in surveys of this kind, is applicable only in high latitudes. In a trigonometrical survey, therefore, of a country lying much farther south than Britain, a different method of ascertaining the longitudes of places must necessarily be adopted.

22. The theorems, which were next proposed to be considered, are those that determine the figure of the earth from the measures of degrees of the curve perpendicular to the meridian, in different latitudes. For this purpose let D′ be a degree of one of these curves, in the latitude φ', and D″ a degree of one of them, in another latitude φ''. Then c being the compression, as before, we have

by § 18, $$mD'=a+c\sin^2\varphi',$$

and also $$mD''=a+c\sin^2\varphi''.$$

Hence $m(D'-D'')=c(\sin^2\varphi'-\sin^2\varphi'')$, and

therefore $$c=\frac{m(D'-D'')}{\sin^2\varphi'-\sin^2\varphi''}.$$

This formula may be rendered more convenient for calculation, by considering that $\sin^2\varphi'=\frac{1-\cos 2\varphi'}{2}$, so that $\sin^2\varphi'-\sin^2\varphi''=\frac{1-\cos 2\varphi'-1+\cos 2\varphi''}{2}=\frac{\cos 2\varphi''-\cos 2\varphi'}{2}$. But $\cos 2\varphi''-\cos 2\varphi'=2\sin(\varphi'+\varphi'')\left(\sin(\varphi'-\varphi'')\right)$, wherefore $\sin^2\varphi'-\sin^2\varphi''=\sin(\varphi'+\varphi'')\left(\sin(\varphi'-\varphi'')\right)$, and $c=\frac{m(D'-D'')}{\sin(\varphi'+\varphi'')\times\sin(\varphi'-\varphi'')}$.

23. In the same manner, because $mD'=a+c\sin^2\varphi'$, by substituting for c, we have

$$mD'=a+\frac{m(D'-D'')\sin^2\varphi}{\sin(\varphi'+\varphi'')\times\sin(\varphi'-\varphi'')},$$ and

$$a=mD'-\frac{m(D'-D'')\sin^2\varphi}{\sin(\varphi'+\varphi'')\times\sin(\varphi'-\varphi'')}.$$

24. Lastly, since $mD'=a+c\sin^2\varphi'$,
and $mD''=a+c\sin^2\varphi''$,
dividing the first of these equations by the second, and rejecting the higher powers of c, we have

$\frac{D'}{D''}=1+\frac{c}{a}(\sin^2\varphi'-\sin^2\varphi'')$, and therefore,

$$\frac{c}{a}=\frac{\frac{D'}{D''}-1}{\sin^2\varphi'-\sin^2\varphi''}.$$ Hence also

$$\frac{c}{a}=\frac{\frac{D'}{D''}-1}{\sin(\varphi'+\varphi'')\times\sin(\varphi'-\varphi'')};$$ or more conveniently for calculation by logarithms,

$$\frac{c}{a}=\frac{D'-D''}{D''\sin(\varphi'+\varphi'')\times\sin(\varphi'-\varphi'')}.$$

25. We may compare this value of $\frac{c}{a}$ with that obtained in § 18, from other data, in order to determine which of the two methods of finding $\frac{c}{a}$ is to be preferred, under given circumstances. Suppose, for instance, a degree of the curve perpendicular to the meridian, in the latitude φ' to be D′, and a degree of the meridian itself in the same latitude to be Δ'; it is required to find in what other latitude φ'', a degree D″, perpendicular to the meridian, must be measured, in order that the comparison of D′ and D″, and of D′ and Δ', may give values of $\frac{c}{a}$, in which the probable error is the same.

Here, agreeably to an observation already made, we may, in order to estimate the error produced in $\frac{c}{a}$, in consequence of an error in the determination of D′ and D″, and Δ', suppose the error to affect $D'-\Delta'$, or $D'-D''$ only, without paying any regard to the variation of D′ in the denominator. Therefore, since by § 18, we have $\frac{c}{a}=\frac{D'-\Delta'}{2D'\cos^2\varphi'}$, and again, by § 24, $\frac{c}{a}=\frac{D'-D''}{D''(\sin^2\varphi'-\sin^2\varphi'')}$, if we suppose equal errors in determining $D'-\Delta'$, and $D'-D''$, and also that these are the only errors, their effect will be the same, in both cases, if $2\cos^2\varphi'=\sin^2\varphi'-$

$\sin^2\varphi''$. Now, if we suppose φ'' the quantity sought, and add $\cos^2\varphi'$ to both sides of the preceding equation, then $3\cos^2\varphi'=\sin^2\varphi'+\cos^2\varphi'-\sin^2\varphi''=1-\sin^2\varphi''=\cos^2\varphi''$. The latitude φ'' therefore must be such, that $\cos\varphi''=\sqrt{3}\times\cos\varphi'$. If, therefore, φ' be such that $\cos\varphi'=\frac{1}{\sqrt{3}}$, the cosine of φ'' will be $=1$ and φ'' therefore $=0$. Now, $54°\ 44'$ is the arch of which the cosine $=\frac{1}{\sqrt{3}}$ nearly, therefore, if a degree of the meridian, and of the perpendicular to it, be measured in latitude $54°\ 44'$, the comparison of these with one another will give a result as accurate as if the degree of the perpendicular, in that latitude, were compared with the degree at the equator, and more accurate of consequence than if any other degree of the perpendicular to the meridian, were to be compared with D′.

26. Hence, also, the comparison of the degree of the meridian, and of the perpendicular to it, in the south of England, is better than if a degree of the perpendicular measured in that latitude were compared with a degree at the equator. For if, in the equation $\cos\varphi''=(\cos\varphi')\sqrt{3}$, we make $\varphi'=50°\ 41'$, (or any thing less than $54°.44'$) φ'' will come out impossible.

27. It may be shown, too, nearly in the same manner, that if a degree of the perpendicular to the meridian were measured in Siberia, as far north as the latitude of $70°$, supposing that to be pos-

sible, and compared with a degree in latitude 45°, or even considerably farther south, it would not give a result so exact as the degree of the meridian and perpendicular measured in the south of England. This shows, that the method of ascertaining the figure of the earth, proposed by the authors of the Trigonometrical Survey, (Phil. Trans. ibid. p. 529,) as a subject of future inquiry, is less exact than that which is founded on their own observations.

28. We may also ascertain, by the same means, the relative accuracy of the method of finding the figure of the earth, from the comparison of a degree of the meridian with a degree of the perpendicular in the same latitude, and of the method of resolving the same problem by the comparison of two degrees of the meridian in different latitudes.

If, then, D be a degree of the meridian, and D′ of the perpendicular, in latitude φ, and if Δ be a degree of the meridian in a different latitude φ', it is required to find whether the most accurate value of $\frac{c}{a}$ will be found, by comparing D and D′, or D and Δ.

Since we have, by what has been already stated, § 4,

$$mD=a-2c+3c\sin^2\varphi, \text{ and}$$

$$m\Delta=a-2c+3c\sin^2\varphi', \text{ we have also}$$

$$\frac{D}{\Delta}=1+\frac{3c}{a}(\sin^2\varphi-\sin^2\varphi') \text{ and therefore,}$$

$$\frac{c}{a}=\frac{D-\Delta}{3\Delta(\sin^2\varphi-\sin^2\varphi')}.$$

Now, it has been already shown, that, by comparing D and D we have $\frac{c}{a}=\frac{D'-D}{2D'\cos^2\varphi}$. Supposing, therefore, equal errors to be committed in the determination of $D-\Delta$, and of $D'-D$, and also paying no regard to the inequality of Δ and D' in the denominators of these fractions, as it is not so great as materially to affect the quantity that is sought for here, we shall have the errors in $\frac{c}{a}$ nearly the same in both formulas, when φ and φ' are such that $2\cos^2\varphi=3\sin^2\varphi-3\sin^2\varphi'$, or when $\frac{2}{3}\cos\varphi^2=\sin^2\varphi-\sin^2\varphi'$, that is, adding $\cos^2\varphi$ to both sides, $\frac{5}{3}\cos^2\varphi=\sin^2\varphi+\cos^2\varphi-\sin^2\varphi'$, and, therefore, $\frac{5}{3}\cos^2\varphi=1-\sin^2\varphi'=\cos^2\varphi'$, or $\cos\varphi'=(\cos\varphi)\sqrt{\frac{5}{3}}$.

29. If, therefore, $\cos\varphi=\sqrt{\frac{3}{5}}$, $\cos\varphi'=1$, that is $\varphi'=0$, so that Δ, the second of the degrees of the meridian, must in this case be under the equator. But $\sqrt{\frac{3}{5}}$ is the cosine of 39°. 14′, in which latitude therefore if D and D′ be measured, the result, by comparing them with one another, is as exact as if D were compared with the degree under the equator. Hence, if D and D′ are measured in a lower latitude than the above, the result will be more exact, than if D were compared with the degree at the equator.

If we suppose D and D′, measured in the south of England, so that $\varphi=50^{\circ}.41$; then we will have $\varphi'=35^{\circ}.7'$, so that D must be compared with a degree of the meridian as far south as $35^{\circ}.7'$, in order that the result may be as good as when D and D are compared with one another.

From this it is evident, that the method of comparing degrees of the meridian, and perpendicular in the same latitude, has even an advantage over the comparison of degrees of the meridian in different latitudes, unless these last are taken at a considerable distance from one another.

In this way may many useful conclusions be derived concerning the degree of credit due to measurements already made, as well as with respect to the selection of the places where they are to be made hereafter. On these I shall enter no further at present, and shall only add, that, besides the advantages or disadvantages which the method of comparing together degrees of the meridian and perpendicular in the same latitude has, and which are subjects of calculation, it has another advantage, which in the case of the British survey is undoubtedly very great, viz. that all the *data* are furnished from one system of trigonometrical operations ; executed according to the same plan, with the same instruments, and by the same observers.

30. One other application of geometrical mea-

surements to discover the figure of the earth yet remains to be considered. This is the comparison of an arch of the meridian with an arch of a parallel of latitude which crosses it. The measure of a parallel of latitude can be executed readily, and is not confined to a small arch as in the case of a perpendicular to the meridian. The plumb-line, while it is carried along the circumference of a parallel to the equator, tends continually to the same point in the earth's axis, so that there is no difficulty in ascertaining the amplitude of the arch measured, providing there be no unusual disturbance of the direction of gravity. As an arch of a parallel to the equator, however, is not the shortest line between two points on the surface of the spheroid, the measurement along that surface will not give the length of the arch truly. To obviate this difficulty, it is only necessary to follow the method so properly introduced into the Trigonometrical Survey, of reducing the measures, both of lines and angles, to the chords and to the planes of the rectilineal triangles contained by them. In this way, the chord of an arch of a parallel of latitude may be determined, however great the arch; and it is worthy of being remarked, that, whatever be the deflections of the plumb-line at the intermediate stations, when the reductions are all properly made, the length of the chord measured will not be affected by them; the amplitude of the arch indeed may be affected by such deflections, if they happen at its extremities; but the effect of this error will be rendered the less, the greater the arch that is measured. We may suppose, therefore, that the chord of a large arch of a parallel of latitude is measured, and the amplitude of the arch itself at the same time accurately ascertained. This last may be done, either by measuring the convergency of the meridians, if it be in a high latitude, or by any other method of ascertaining differences of longitude which admits of great accuracy. The chord being thus given in fathoms, and the arch subtended by it being given in degrees and minutes, the radius of the parallel itself becomes known.

31. Now, if we would compare the radius of a parallel thus found, with a large arch of the meridian, we shall have by that means a determination of the figure of the earth, not less to be relied on than that given in the beginning of this paper. The investigation is easy by help of the theorems in § 5 and 6. Let FO be the radius of a parallel to the equator, which passes through F, the latitude of which is φ, and is supposed known; and let FO found by the method just described be $=r$, then, as in § 4, $r=\frac{a^2\cos\varphi}{\sqrt{a^2\cos\varphi^2+b^2\sin\varphi^2}}=$ $\frac{a\cos\varphi^2}{a\sqrt{1-\frac{2c}{a}\sin\varphi^2}}$, according to the method of reduc-

tion followed in the preceding articles of this paper. Then, because $\sqrt{1-\frac{2c}{a}\sin\varphi^2}=1+\frac{c}{a}\sin\varphi^2$ nearly, we have $r=\cos\varphi(1+\frac{c}{a}\sin\varphi^2)=a\cos\varphi+c\sin\varphi^2\cos\varphi$, or if we divide by $\cos\varphi$, $\frac{r}{\cos\varphi}=a+c\sin\varphi^2$. Let $\frac{r}{\cos\varphi}=l$, then $l=a+c\sin\varphi^2$.

32. Again, if φ' and φ'' are the latitudes of the extremities of an arch of the meridian, the length of which has been measured, and found $=l$, then, according to § 5, we have $l'=a(\varphi''-\varphi')-\frac{c}{2}\left((\varphi''-\varphi')+\frac{3}{2}(\sin2\varphi''-\sin2\varphi')\right)$. If, therefore, m be the coefficient of a, in the former equation, and n the coefficient of c; and if m' be the coefficient of a, in the latter equation, and n' of c, we have, as in § 6,

$$a=\frac{n'l-nl'}{mn'-m'n},\text{ and }c=\frac{m'l-ml'}{mn'-m'n},\text{ or since }m=1,$$

$$a=\frac{n'l-nl'}{n'-m'n},\text{ and }c=\frac{m'l-l'}{n'-m'n}\text{ ; also }\frac{c}{a}=\frac{m'l-l'}{n'l-nl'}.$$

33. In this way of determining a and c, the parallel of latitude may either intersect the arch of the meridian measured or not. If it intersect that arch, this method may have the same advantage that was taken notice of in another solution, viz. that the whole of the *data* may be furnished from the same system of trigonometrical operations. Thus, in the survey of Great Britain, an arch of 5 or 6 degrees of a parallel to the equator might be

measured, and compared with the whole length of the meridian, comprehended between the northern and southern extremities of the Island, amounting nearly to 9 degrees.

It is plain, from what has already been said, that the result deduced from this comparison would possess every advantage, and would be entitled to more credit, than any determination of the figure of the earth that is yet known.

34. On the supposition that, in the survey of a country, the measurement is made along a series of triangular planes, all given in position and magnitude, there is yet another method of determining the figure of the earth, more general than any of the former. On the supposition just mentioned, it is evident, that the length of a straight line, or chord, drawn from a given angle of any one of these triangles, to a given angle of any other of them, may be found by trigonometrical calculation. Let the latitudes be observed at the extremities of this chord, and also the difference of longitude; then, from the nature of an ellipsoid, the length of this same chord may be expressed, in terms of the axes a and b, together with the latitudes of the extremities of the chord, and the difference of longitude between them; and this expression being put equal to the length of the chord measured, will give an equation, in which all the quantities are known, except a and b. Further, if $a = b + c$, and if the said

expression be reduced into a series, with the powers of c ascending, that series will converge very rapidly, because c is small in respect of a; then, for a first approximation, we may reject all the terms that involve the powers of c higher than the first, by which means we shall have a simple equation of the form $ma+nc=l$, where m and n are functions of the latitudes and difference of longitude, and l is the length of the chord.

Now, if a similar equation be derived from the measurement of any other chord, these two equations will give a and c in the same manner as in § 6; and thus, from the measurement of any two chords, the figure of the earth will be determined.

35. The length of the chords, thus measured, should be great, so that they may, if possible, subtend angles of several degrees, and their position will be most favourable when one of them is in the plane of the meridian, and the other nearly at right angles to it. The numerical computation will be found less laborious than might be imagined; but the complete solution of the problem, and the full detail of the investigation, I am under the necessity of delaying to some future communication.

There seems to be but one difficulty of any consequence that stands in the way of this method of determining the figure of the earth. It arises from this, that the ascertaining the position of the supposed series of triangular planes relatively to one another, involves in it the allowance to be made for the terrestrial refraction, which it must be confessed is not accurately known, and is the more difficult to determine, that it is unavoidably combined with the irregularities in the direction of gravity. It is possible, indeed, to separate these two sources of error, but not without a system of experiments instituted directly for that purpose.

36. The determination of the difference of longitude, which enters necessarily into this problem, except in the case when both chords are in the direction of the meridian, must also be performed with great accuracy. Among the different ways of doing this, that which proceeds by observing the convergency of the meridians, though the best accommodated to the nature of a trigonometrical survey, is not the least liable to objection. For, not to mention that it is only practicable in high latitudes, we must observe, that it always implies a correction on account of the ellipticity of the meridian, which is therefore necessarily hypothetical, and depends on the very thing that is to be found. This inconvenience, however, may be obviated by repeated approximations, and by an accurate solution of *spheroidal* triangles. On this latter subject it was my intention to offer to the Society some theorems, that contain more direct and fuller rules for this kind of trigonometry than any that I have

yet met with. I am under the necessity, however, of reserving these, as well as the solution of the problem above mentioned, for the subjects of some future communication. In the mean time, I think it is material to observe, that the principle laid down by Mr Dalby, viz. that in a spheroidal triangle, of which the angle at the pole and the two sides are given, the sum of the angles at the base is the same as in a spherical triangle, having the same sides, and the same vertical angle, is not strictly true, unless the eccentricity of the spheroïd be infinitely small, or the triangle be very nearly isosceles. The application of the principle may therefore lead into error, unless it be made with due attention to these restrictions. The gentleman just named will forgive a remark, which I certainly should not have made, if I had been less interested for the success of the work, in which he has assisted with so much ability.

FINIS.

ON

THE SOLIDS

OF

GREATEST ATTRACTION.

ON

THE SOLIDS

OF

GREATEST ATTRACTION.*

The investigations which I have at present the honour of submitting to the Royal Society, were suggested by the experiments which have been made of late years concerning the gravitation of terrestrial bodies, first, by Dr Maskelyne, on the Attraction of Mountains, and afterwards by Mr Cavendish, on the Attraction of Leaden Balls.

In reflecting on these experiments, a question naturally enough occurred, what figure ought a given mass of matter to have, in order that it may attract a particle in a given direction, with the greatest force possible? This seemed an inquiry not of mere curiosity, but one that might be of use in the further prosecution of such experi-

* From the Transactions of the Royal Society of Edinurgh, Vol. VI. (1809.)—Ed.

ments as are now referred to. On considering the question more nearly, I soon found, though it belongs to a class of problems of considerable difficulty, which the Calculus Variationum is usually employed to resolve, that it nevertheless admits of an easy solution, and one leading to results of remarkable simplicity, such as may interest Mathematicians by that circumstance, as well as by their connection with experimental inquiries.

In the problem thus proposed, no condition was joined to that of the greatest attraction, but that of the quantity of homogeneous matter being given. This is the most general state of the problem. It is evident, however, that other conditions may be combined with the two preceding; it may be required that the body shall have a certain figure, conical, for example, cylindric, &c. and the problem, under such restrictions, may be still more readily applicable to experiments than in its most general form.

Though the question, thus limited, belongs to the common method of Maxima and Minima, it leads to investigations that are in reality considerably more difficult than when it is proposed in its utmost generality.

Among the following investigations, there are also some that have a particular reference to the experiments on Schehallien. A few years ago, an attempt was made by Lord Webb Seymour and myself, toward such a survey of the rocks which compose that mountain, as might afford a tolerable estimate of their specific gravity, and thereby serve to correct the conclusions, deduced from Dr Maskelyne's observations, concerning the mean density of the earth. The account of this survey, and of the conclusions arising from it, belongs naturally to the Society under whose direction the original experiment was made; what is offered here, is an investigation of some of the theorems employed in obtaining those conclusions. When a new element, the heterogeneity of the mass, or the unequal distribution of density in the mountain, was to be introduced into the calculations, the ingenious methods employed by Dr Hutton could not always be pursued. The propositions that relate to the attraction of a half, or quarter cylinder, on a particle placed in its axis, are intended to remedy this inconvenience, and will probably be found of use in all inquiries concerning the disturbance of the direction of the plumb-line by inequalities, whether in the figure or density of the exterior crust of the globe.

The first of the problems here resolved, has been treated of by Boscovich; and his solution is mentioned in the catalogue of his works, as published in the Memoirs of a Philosophical Society at Pisa. I have never, however, been able to procure a sight of these memoirs, nor to obtain any account of the solution just mentioned, and therefore am sensible of hazarding a good deal, when I treat of a subject that has passed through the hands of so able a mathematician, without knowing the conclusions which he has come to, or the principles which he has employed in his investigation. In such circumstances, if my result is just, I cannot reasonably expect it to be new; and I should, indeed, be much alarmed to be told, that it has not been anticipated. The other problems contained in this paper, as far as I know, have never been considered.

1. To find the solid into which a mass of homogeneous matter must be formed, in order to attract a particle given in position, with the greatest force possible, in a given direction.

Let A (Fig. 21) be the particle given in position, AB the direction in which it is to be attracted; and ACBH a section of the solid required, by a plane passing through AB.

Since the attraction of the solid is a maximum, by hypothesis, any small variation in the figure of the solid, provided the quantity of matter remain the same, will not change the attraction in the direction AB. If, therefore, a small portion of matter be taken from any point C, in the superficies of the solid, and placed at D, another point in the same superficies, there will be no variation produced in the force which the solid exerts on the particle A, in the direction AB.

The curve ACB, therefore, is the locus of all the points in which a body being placed, will attract the particle A in the direction AB, with the same force.

This condition is sufficient to determine the nature of the curve ACB. From C, any point in that curve, draw CE perpendicular to AB; then if a mass of matter placed at C be called m^3, $\frac{m^3}{AC^2}$ will be the attraction of that mass on A, in the direction AC, and $\frac{m^3 \times AE}{AC^3}$ will be its attraction in the direction AB. As this is constant, it will be equal to $\frac{m^3}{AB^2}$, and therefore $AB^2 \times AE = AC^3$.

All the sections of the required solid, therefore, by planes passing through AB, have this property, that $AC^3 = AB^2 \times AE$; and as this equation is sufficient to determine the nature of the curve to which it belongs, therefore all the sections of the solid, by planes that pass through AB, are similar and equal curves; and the solid of consequence may be conceived to be generated by the revolution of ACB, any one of these curves, about AB as an axis.

The solid so generated may be called the *Solid*

of greatest Attraction; and the line ACB, the *Curve of equal Attraction.*

2. To find the equation between the co-ordinates of ACB, the curve of equal attraction.

From C (Fig 21.) draw CE perpendicular to AB; let $AB=a$, $AE=x$, $EC=y$. We have found $AB^2 \times AE = AC^3$, that is, $a^2 x = (x^2+y^2)^{\frac{3}{2}}$, or $a^4x^2 = (x^2+y^2)^3$, which is an equation to a line of the 6th order.

To have y in terms of x, $x^2+y^2 = a^{\frac{4}{3}} x^{\frac{2}{3}}$, $y^2 = a^{\frac{4}{3}} x^{\frac{2}{3}} - x^2$, and $y = x^{\frac{1}{3}} \sqrt{a^{\frac{4}{3}} - x^{\frac{4}{3}}}$.

Hence $y=0$, both when $x=0$, and when $x=a$. Also if x be supposed greater than a, y is impossible. No part of the curve, therefore, lies beyond B.

The parts of the curve on opposite sides of the line AB, are similar and equal, because the positive and negative values of y are equal. There is also another part of the curve on the side of A, opposite to B, similar and equal to ACB; for the values of y are the same whether x be positive or negative.

3. The curve may easily be constructed without having recourse to the value of y just obtained.

Let $AB=a$, (Fig 21,) $AC=z$, and the angle $BAC=\varphi$. Then $AE = AC \times \cos\varphi = z\cos\varphi$, and so $a^2 z\cos\varphi = z^3$, or $a^2\cos\varphi = z^2$; hence $z = a\sqrt{\cos\varphi}$.

From this formula a value of AC or z may be ound, if φ or the angle BAC be given; and if it be required to find z in numbers, it may be conveniently calculated from this expression. A geometrical construction may also be easily derived from it. For if with the radius AB, a circle BFH be described from the centre A; if AC be produced to meet the circumference in F, and if FG be drawn at right angles to AB, then $\frac{\mathrm{AG}}{\mathrm{AB}}=\cos\varphi$, and so $z=a\sqrt{\frac{\mathrm{AG}}{\mathrm{AB}}}=\sqrt{\mathrm{AB}\times\mathrm{AG}}=\mathrm{AC}$.

Therefore, if from the centre A, with the distance AB, a circle BFH be described, and if a circle be also described on the diameter AB, as AKB, then drawing any line AF from A, meeting the circle BFH in F, and from F letting fall FG perpendicular on AB, intersecting the semicircle AKB in K; if AK be joined, and AC made equal to AK, the point C is in the curve.

For $\mathrm{AK}=\sqrt{\mathrm{AB}\times\mathrm{AG}}$, from the nature of the semicircle, and therefore $\mathrm{AC}=\sqrt{\mathrm{AB}\times\mathrm{AG}}$, which has been shown to be a property of the curve. In this way, any number of points of the curve may be determined; and the *Solid of greatest Attraction* will be described, as already explained, by the revolution of this curve about the axis AB.

4. To find the area of the curve ACB.

1. Let ACE, AFG (Fig. 22,) be two radii, in-

definitely near to one another, meeting the curve ACB in C and F, and the circle, described with the radius AB, in E and G. Let AC$=z$ as before, the angle BAC$=\varphi$, and AB$=a$. Then GE$=a\dot{\varphi}$, and the area AGE$=\frac{1}{2}a^2\dot{\varphi}$, and since AE^2 : AC^2 : : Sect. AEG : Sect. ACF, the sector ACF$=\frac{1}{2}z^2\dot{\varphi}$. But $z^2=a^2\cos\varphi$, (§ 3,) whence the sector ACF, or the fluxion of the area ABC$=\frac{1}{2}a^2\dot{\varphi}\cos\varphi$, and consequently the area ABC$=\frac{1}{2}a^2\sin\varphi$, to which no constant quantity need be added, because it vanishes when $\varphi=0$, or when the area ABC vanishes.

The whole area of the curve, therefore, is $\frac{1}{2}a^2$, or $\frac{1}{2}$ AB^2; for when φ is a right angle $\sin\varphi=1$. Hence the area of the curve on both sides of AB is equal to the square of AB.

2. The value of x, when y is a maximum, is easily found. For when y, and therefore y^2 is a maximum, $\frac{2}{3}a^{\frac{1}{3}}x^{-\frac{1}{3}}=2x$, or $3x^{\frac{4}{3}}=a^{\frac{4}{3}}$, that is,

$$x=\frac{a}{3^{\frac{3}{4}}}=\frac{a}{\sqrt[4]{27}}.$$

Hence, calling b the value of y when a maximum, $b^2=a^{\frac{4}{3}}\frac{a^{\frac{2}{3}}}{27^{\frac{1}{6}}}-\frac{a^2}{27^{\frac{1}{2}}}=a^2\left(\frac{27^{\frac{1}{3}}-1}{27^{\frac{1}{2}}}\right)=\frac{2a^2}{\sqrt{27}}$, and so $b=a\frac{\sqrt{2}}{\sqrt[4]{27}}$, and therefore $a:b::\sqrt[4]{27}:\sqrt{2}$, or as 11 : 7 nearly.

3. It is material to observe, that the radius of curvature at A is infinite. For since $y^2=a^{\frac{4}{3}}x^{\frac{2}{3}}-x^2$, $\frac{y^2}{x}=\frac{a^{\frac{4}{3}}}{x^{\frac{1}{3}}}-x$. But when x is very small, or y indefinitely near to A, $\frac{y^2}{x}$ becomes the diameter of the circle having the same curvature with ACB at A, and when x vanishes, this value of $\frac{y^2}{x}$, or $\frac{a^{\frac{4}{3}}}{x^{\frac{1}{3}}}-x$, becomes infinite, because of the divisor $x^{\frac{1}{3}}$ being in that case $=0$. The diameter, therefore, and the radius of curvature at A are infinite. In other words, no circle, having its centre in AB produced, and passing through A, can be described with so great a radius, but that, at the point A, it will be within the curve of equal attraction.

The solid of greatest attraction, then, at the extremity of its axis, where the attracted particle is placed, is exceedingly flat, approaching more nearly to a plane than the superficies of any sphere can do, however great its radius.

4. To find the radius of curvature at B, the other extremity of the axis, since $y^2=a^{\frac{4}{3}}x^{\frac{2}{3}}-x^2$, if we divide by $a-x$, we have $\frac{y^2}{a-x}=\frac{a^{\frac{4}{3}}x^{\frac{2}{3}}-x^2}{a-x}$. But at B, when $a-x$, or the abscissa reckoned from B vanishes, $\frac{y^2}{a-x}$ is the diameter of the circle having

the same curvature with ACB in B. But when $a-x=0$, or $a=x$, both the numerator and denominator of the fraction $\frac{a^{\frac{4}{3}}x^{\frac{2}{3}}-x^2}{a-x}$ vanish, so that its ultimate value does not appear. To remove this difficulty, let $a-x=z$, or $x=a-z$, then we have $y^2=a^{\frac{4}{3}}(a-z)^{\frac{2}{3}}-(a-z)^2$. But when z is extremely small, its powers, higher than the first, may be rejected; and therefore $(a-z)^{\frac{2}{3}}=a^{\frac{2}{3}}\left(1-\frac{z}{a}\right)^{\frac{2}{3}}=a^{\frac{2}{3}}$ $(1-\frac{2z}{3a}$, &c.). Therefore the equation to the curve becomes in this case, $y=a^{\frac{4}{3}}a^{\frac{2}{3}}\left(1-\frac{2z}{3a}\right)-a^2+2az$ $=a^2-\frac{2}{3}az-a^2+2az=\frac{4}{3}az.$

Hence $\frac{y^2}{2z}$, or the radius of curvature at B $=\frac{2}{3}a$. The curve, therefore, at B falls wholly without the circle BKA, described on the diameter AB, as its radius of curvature is greater. This is also evident from the construction.

5. To find the force with which the solid above defined attracts the particle A in the direction AB.

Let b (Fig. 22,) be a point indefinitely near to B, and let the curve Acb be described similar to ACB. Through C draw CcD perpendicular to AB, and suppose the figure thus constructed to revolve about AB; then each of the curves ACB,

A*cb* will generate a solid of greatest attraction; and the excess of the one of these solids above the other, will be an indefinitely thin shell, the attraction of which is the variation of the attraction of the solid ACB, when it changes into A*cb*.

Again, by the line DC, when it revolves along with the rest of the figure about AB, a circle will be described; and by the part C*c*, a circular ring, on which, if we suppose a solid of indefinitely small altitude to be constituted, it will make the element of the solid shell AC*c*. Now the attraction exerted by this circular ring upon A, will be the same as if all the matter of it were united in the point C, and the same, therefore, as if it were all united in B.

But the circular ring generated by C*c*, is $=\pi(DC^2-Dc^2)=2\pi DC\times Cc$. Now $2DC\times Cc$ is the variation of y^2, or DC^2, while DC passes into D*c*, and the curve BCA into the curve *bc*A; that is $2DC\times Cc$ is the fluxion of y^2, or of $a^{\frac{4}{3}}x^{\frac{2}{3}}-x^2$, taken on the supposition that x is constant and a variable, viz. $\frac{4}{3}a^{\frac{1}{3}}\dot{a}x^{\frac{2}{3}}$. Therefore the space generated by

$$Cc=\frac{4\pi}{3}a^{\frac{1}{3}}x^{\frac{2}{3}}\dot{a}.$$

If this expression be multiplied by x, we have the element of the shell $=\frac{4\pi}{3}a^{\frac{1}{3}}x^{\frac{2}{3}}\dot{a}\dot{x}$.

In order to have the solidity of the shell AC B*bc*, the above expression must be integrated relatively to x, that is, supposing only x variable, and it is then $\frac{3}{5}\left(\frac{4\pi}{3}a^{\frac{1}{3}}x^{\frac{5}{3}}\dot{a}\right)+C$. But $C=0$, because the fluent vanishes when x vanishes, therefore the portion of the shell $ACc = \frac{4}{5}x^{\frac{5}{3}}a^{\frac{1}{3}}\dot{a}$, and when $x=a$, the whole shell $= \frac{4\pi}{5}a^2\dot{a}$.

Now, if the whole quantity of matter in the shell were united at B, its attractive force exerted on A, would be the same with that of the shell; therefore the whole force of the shell $=\frac{4\pi}{5}\dot{a}$. The same is true for every other indefinitely thin shell into which the solid may be supposed to be divided; and therefore the whole attraction of the solid is equal to $\int\frac{4\pi}{5}\dot{a}$, supposing a variable, that is $=\frac{4\pi}{5}a$.

Hence we may compare the attraction of this solid with that of a sphere of which the axis is AB, for the attraction of that sphere $=\frac{\pi}{6}a^3\frac{4}{a^2}=\frac{2\pi}{3}a$. The attraction of the solid ADBH′, (Fig. 21,) is therefore to that of the sphere on the same axis as $\frac{4\pi}{5}a$ to $\frac{2\pi}{3}a$, or as 6 to 5.

6. To find the content of the solid ADBH′, we need only integrate the fluxionary expression for the content of the shell, viz. $\frac{4\pi}{5}a^2\dot{a}$. We have then

$\frac{4\pi}{15}a^3 =$ the content of the solid ADBH′. Since the solidity of the sphere on the axis a is $= \frac{\pi}{6}a^3$, the content of the solid ADBH′ is to that of the sphere on the same axis as $\frac{4\pi}{15}a^3$ to $\frac{\pi}{6}a^3$; that is, as $\frac{4}{15}$ to $\frac{1}{6}$, or as 8 to 5.

7. Lastly, To compare the attraction of this solid with the attraction of a sphere of equal bulk, let $m^3 =$ any given mass of matter formed into the solid ADBH′; then for determining AB, we have this equation, $\frac{4\pi}{15}a^3 = m^3$, and $a = m\sqrt[3]{\frac{15}{4\pi}}$; and therefore also the attraction of the solid, (which is $\frac{4\pi}{5}a$)

$$= \frac{4\pi}{5}m\sqrt[3]{\frac{15}{4\pi}} = m\left(\frac{4.5^{\frac{1}{3}}.3^{\frac{1}{3}}.\pi^{\frac{2}{3}}}{5.4^{\frac{1}{3}}}\right) = m\left(\frac{4^{\frac{2}{3}}.3^{\frac{1}{3}}.\pi^{\frac{2}{3}}}{5^{\frac{2}{3}}}\right) = m\sqrt[3]{\frac{48\pi^2}{25}}:$$

Again, if m^3 be formed into a sphere, the radius of that sphere $= m\sqrt[3]{\frac{3}{4\pi}}$, and the attraction of it on a particle at its surface $= \frac{m^3}{m^2\left(\frac{3}{4\pi}\right)^{\frac{2}{3}}} = m\frac{(16\pi^2)^{\frac{1}{3}}}{9^{\frac{1}{3}}}$.

Hence the attraction of the solid ADBH′, is to that of a sphere equal to it, as $m\left(\frac{48}{25}\pi^2\right)^{\frac{1}{3}}$ to m

$\left(\frac{16\pi^2}{9}\right)^{\frac{1}{3}}$.; that is, as $(27)^{\frac{1}{3}}$ to $(25)^{\frac{1}{3}}$, or as 3 to the cube-root of 25.

The ratio of 3 to $\sqrt[3]{25}$, is nearly that of 3 to $3-\frac{2}{27}$, or of 81 to 79; and this is therefore also nearly equal to the ratio of the attraction of the solid ADBH to that of a sphere of equal magnitude.

8. It has been supposed in the preceding investigation, that the particle on which the solid of greatest attraction exerts its force is in contact with that solid. Let it now be supposed, that the distance between the solid and the particle is given; the solid being on one side of a plane, and the particle at a given distance from the same plane on the opposite side. The mass of matter which is to compose the solid being given, it is required to construct the solid.

Let the particle to be attracted be at A (Fig. 23,) from A draw AA′ perpendicular to the given plane, and let EF be any straight line in that plane, drawn through the point A′; it is evident that the axis of the solid required must be in AA′ produced. Let B be the vertex of the solid, then it will be demonstrated as has been done above, that this solid is generated by the revolution of the curve of *equal attraction*, (that of which the equation is $y^2=a^{\frac{4}{3}}x^{\frac{2}{3}}-x^2$), about the axis of which one extremity is at A, and of which the length must be found from the quantity of matter in the solid.

The solid required, then, is a segment of the solid of greatest attraction, having B for its vertex, and a circle, of which A′E or A′F is the radius, for its base.

To find the solid content of such a segment, CD being $=y$, and $AC=x$, we have $y^2=a^{\frac{4}{3}}x^{\frac{2}{3}}-x^2$, and $\pi y^2\dot{x}=\pi a^{\frac{4}{3}}x^{\frac{2}{3}}\dot{x}-\pi x^2\dot{x}=$ the cylinder which is the element of the solid segment.

Therefore $\int\pi y^2\dot{x}$, or the solid segment intercepted between B and D must be $\frac{3}{5}\pi a^{\frac{4}{3}}x^{\frac{5}{3}}-\frac{1}{3}\pi x^3+C$. This must vanish when $x=a$, or when C comes to B, and therefore $C=-\frac{4\pi}{15}a^3$. The segment, therefore, intercepted between B and C, the line AC being x, is $\frac{4\pi}{15}a^3-\frac{3\pi}{5}a^{\frac{4}{3}}x^{\frac{5}{3}}+\frac{\pi}{3}x^3$.

This also gives $\frac{4\pi}{15}a^3$, for the content of the whole solid, when $x=0$, the same value that was found by another method at § 6.

Now, if we suppose x to be $=AA'$, and to be given $=b$, the solid content of the segment becomes $\frac{4\pi}{15}a^3-\frac{3}{5}\pi a^{\frac{4}{3}}b^{\frac{5}{3}}+\frac{\pi}{3}b^3$, which must be made equal to the given solidity, which we shall suppose $=m^3$, and from this equation, a which is yet unknown, is to be determined. If then, for $a^{\frac{1}{3}}$ we put u, we have $\pi\left(\frac{4}{15}u^9-\frac{3}{5}b^{\frac{5}{3}}u^4+\frac{1}{3}b^3\right)=m^3$, or $\frac{4}{15}u^9-\frac{3}{5}b^{\frac{5}{3}}u^4=\frac{m^3}{\pi}-\frac{1}{3}b^3$ and $u^9-\frac{9}{4}b^{\frac{5}{3}}u^4=\frac{15m^3}{4\pi}-\frac{15}{12}b^3$.

The simplest way of resolving this equation, would be by the rule of false position. In some particular cases, it may be resolved more easily; thus, if $\frac{15m^3}{\pi}-\frac{15}{12}b^3=0$, $u^9-\frac{9}{4}b^{\frac{5}{3}}u^4=0$, and $u^5=\frac{9}{4}b^{\frac{5}{3}}$, that is $a^{\frac{5}{3}}=\frac{9}{4}b^{\frac{5}{3}}$ or $a=b\left(\frac{9}{4}\right)^{\frac{3}{5}}=b\sqrt[5]{\frac{729}{64}}$.

9. If it be required to find the equation to the superficies of the solid of greatest attraction, and also to the sections of it parallel to any plane passing through the axis; this can readily be done by help of what has been demonstrated above.

1. Let AHB (Fig. 24,) be a section of the solid, by a plane through AB its axis. Let G be any point in the superficies of the solid, GF a perpendicular from G on the plane AHB, and FE a perpendicular from F on the axis. Let $AE=x$, $EF=z$, $FG=v$, then x, z, and v are the three coordinates by which the superficies is to be defined. Let $AB=a$, $EH=y$, then, from the nature of the curve AHB, $y^2=a^{\frac{4}{3}}x^{\frac{2}{3}}-x^2$. But because the plane GEH is at right angles to AB, G and H are in the circumference of a circle of which E is the centre; so that $GE=EH=y$. Therefore $EF^2+FG^2=EH^2$, that is, $z^2+v^2=y^2$, and by substitution

for y^2 in the former equation, $z^2+v^2=a^{\frac{4}{3}}x^{\frac{2}{3}}-x^2$, or $(x^2+z^2+v^2)^3=a^4x^2$, which is the equation to the superficies of the solid of greatest attraction.

2. If we suppose EF, that is z, to be given $=b$, and the solid to be cut by a plane through FG and CD, (CD being parallel to AB,) making on the surface of the solid the section DGC; and if AK be drawn at right angles to AB, meeting DC in K, then we have, by writing b for z in either of the preceding equations, $b^2+v^2=a^{\frac{4}{3}}x^{\frac{2}{3}}-x^2$, and $v^2=a^{\frac{4}{3}}x^{\frac{2}{3}}-x^2-b^2$ for the equation of the curve DGC, the coordinates being GF and FK, because FK is equal to AE or x.

This equation also belongs to a curve of equal attraction; the plane in which that curve is being parallel to AB, the line in which the attraction is estimated, and distant from it by the space b.

Instead of reckoning the abscissa from K, it may be made to begin at C. If AL or CK$=h$, then the value of h is determined from the equation, $b^2=a^{\frac{4}{3}}h^{\frac{2}{3}}-h^2$, and if $x=h+u$, u being put for CF, $v^2=a^{\frac{4}{3}}(h+u)^{\frac{2}{3}}-(h+u)^2-a^{\frac{4}{3}}h^{\frac{2}{3}}+h^2$, or $v^2+(h+u)^2+b^2=a^{\frac{4}{3}}(h+u)^{\frac{2}{3}}$, or $(v^2+(h+u)^2+b^2)^3=a^4(h+u)^2$.

When b is equal to the maximum value of the ordinate EH, (§ 4. 2,) the curve CGD goes away into a point; and if b be supposed greater than this, the equation to the curve is impossible.

10. The solid of greatest attraction may be found, and its properties investigated, in the way that has now been exemplified, whatever be the law of the attracting force. It will be sufficient, in any case, to find the equation of the generating curve, or the curve of equal attraction.

Thus, if the attraction which the particle C (fig. 21,) exerts on the given particle at A, be inversely as the m power of the distance, or as $\frac{1}{\text{AC}^m}$, then the attraction in the direction AE will be $\frac{\text{AE}}{\text{AC}^{m+1}}$, and if we make this $=\frac{1}{\text{AB}^m}$, we have $\frac{\text{AE}}{\text{AC}^{m+1}}=\frac{1}{\text{AB}^m}$, or making AE$=x$, EC$=y$, and AB$=a$, as before, $\frac{x}{(x^2+y^2)^{\frac{m+1}{2}}}=\frac{1}{a^m}$, or $a^m x=(x^2+y^2)^{\frac{m+1}{2}}$, and $x^2+y^2=a^{\frac{2m}{m+1}}x^{\frac{2}{m+1}}$, or $y^2=a^{\frac{2m}{m+1}}x^{\frac{2}{m+1}}-x^2$.

If $m=1$, or $m+1=2$, this equation becomes $y^2=ax-x^2$, being that of a circle of which the diameter is AB. If, therefore, the attracting force were inversely as the distance, the solid of greatest attraction would be a sphere.

If the force be inversely as the cube of the distance, or $m=3$, and $m+1=4$, the equation is $y^2=a^{\frac{3}{2}}x^{\frac{1}{2}}-x^2$, which belongs to a line of the 4th order.

If $m=4$, and $m+1=5$, the equation is $y^2=a^{\frac{8}{5}}x^{\frac{2}{5}}-x^2$; which belongs to a line of the 10th order.

In general, if m be an even number, the order of the curve is $(m+1)2$; but if m be an odd number, it is $m+1$ simply.

11. In the same manner that the solid of greatest attraction has been found, may a great class of similar problems be resolved. Whenever the property that is to exist in its greatest or least degree, belongs to all the points of a plane figure, or to all the points of a solid, given in magnitude, the question is reduced to the determination of the locus of a certain equation, just as in the preceding example.

Let it, for instance, be required to find a solid given in magnitude, such, that from all the points in it, straight lines being drawn to any assigned number of given points, the sum of the squares of all the lines so drawn shall be a minimum. It will be found, by reasoning as in the case of the solid of greatest attraction, that the superficies bounding the required solid must be such that the sum of the squares of the lines drawn from any point in it, to all the given points, must be always of the same magnitude. Now, the sum of the squares of the lines drawn from any point to all the given points, may be shown by plane geometry to be equal to the square of the line drawn to the centre of gravity of these given points, multiplied by the number of points, together with a given space. The line, therefore, drawn from any point in the required superficies to the centre of gravity of the given points, is given in magnitude, and therefore the superficies is that of a sphere, having for its centre the centre of gravity of the given points.

The magnitude of the sphere is next determined from the condition, that its solidity is given.

In general, if x, y, and z, are three rectangular co-ordinates that determine the position of any point of a solid given in magnitude, and if the value of a certain function Q, of x, y, and z, be computed for each point of the solid, and if the sum of all these values of Q added together, be a maximum or a minimum, the solid is bounded by a superficies in which the function Q is every where of the same magnitude. That is, if the triple integral $\int \dot{x} \int \dot{y} \int Q\dot{z}$ be the greatest or least possible, the superficies bounding the solid is such that $Q=A$, a constant quantity.

The same holds of plane figures; the proposition is then simpler, as there are only two co-ordinates, so that $\int \dot{x} \int Q\dot{y}$ is the quantity that is to be a maximum or a minimum, and the line bounding the figure is defined by the equation $Q=A$.

All the questions, therefore, which come under

this description, though they belong to an order of problems which requires, in general, the application of one of the most refined inventions of the New Geometry, the Calculus Variationum, form a particular division admitting of resolution by much simpler means, and directly reducible to the construction of loci.

In these problems, also, the synthetical demonstration will be found extremely simple. In the instance of the solid of greatest attraction, this holds remarkably. Thus, it is obvious, that (fig. 21,) any particle of matter placed without the curve ADBH, will attract the particle at A in the direction AB, less than any of the particles in that curve, and that any particle of matter within the curve, will attract the particle at A more than any particle in the curve, and more, *à fortiori*, than any particle without the curve. The same is true of the whole superficies of the solid. Now, if the figure of the solid be any how changed, while its quantity of matter remains the same, as much matter must be expelled from within the surface, at some one place C, as is accumulated without the surface at some other point H′. But the action of any quantity of matter within the superficies ADBH′ on A, is greater than the action of the same without the superficies ADBH′. The solid ADBH′ therefore, by any change of its figure, must lose more attraction than it gains; that is, its attraction is diminished by every such change, and therefore it is itself the solid of greatest attraction. Q. E. D.

12. The preceding theorems relate to the solids, which, of all solids whatsoever of a given content, have the greatest attraction in a given direction. It may be interesting also to know, among bodies of a given kind, and a given solid content, for example, among cones, cylinders, or parallelepipeds, given in magnitude, which has the greatest attractive power, in the direction of a certain straight line. We shall begin with the cone.

Let ABC (fig. 25,) be a cone of which the axis is AD, required to find the angle BAC, when the force which the cone exerts, in the direction AD, on the particle A at its vertex, is greater than that which any other cone, of the same solid content, can exert in the direction of its axis, on a particle at its vertex.

It is known, if π be the semicircumference of the circle of which the radius is 1, that is, if $\pi=3.14159$, &c. that the attraction of the cone ABC, on the particle A, in the direction AD, is $=2\pi\left(\text{AD}-\frac{\text{AD}^2}{\text{AB}}\right)$. (Simpson's Fluxions, Vol. II. Art. 377.)

Let AD$=x$, AB$=z$, the solid content of the cone $=m^3$, and its attraction $=$A.

Then $\text{A}=2\pi\left(x-\frac{x^2}{z}\right)$, and $\pi x(z^2-x^2)=3m$

The quantity $x-\frac{x^2}{z}$, is to be a maximum, and therefore, $\dot{x}-\frac{2xz\dot{x}-x^2\dot{z}}{z^2}=0$, or $z^2-2xz+x^2\left(\frac{\dot{z}}{\dot{x}}\right)=0$.

Again, from the equation $\pi x(z^2-x^2)=3m^3$, we have $2xz\dot{z}+z^2\dot{x}-3x^2\dot{x}=0$, and $\frac{\dot{z}}{\dot{x}}=\frac{3x^2-z^2}{2xz}$, and by substituting this value of $\frac{\dot{z}}{\dot{x}}$ in the former equation, we have $z^2-\frac{5}{2}xz+\frac{3}{2}\left(\frac{x^3}{z}\right)=0$.

As this equation is homogeneous, if we make $\frac{x}{z}=u$, we will obtain an equation involving u only, and therefore determining the ratio of z to x, or of AB to AD. Substituting, accordingly, uz for x in the last equation, we have $z^2-\frac{5}{2}uz^2+\frac{3}{2}u^3z^2=0$, and $1-\frac{5}{2}u+\frac{3}{2}u^3=0$.

This equation is obviously divisible by $u-1$, and when so divided, gives $\frac{3}{2}u^2+\frac{3}{2}u-1=0$, or $u^2+u=\frac{2}{3}$, whence $u=-\frac{1}{2}\pm\sqrt{\frac{11}{12}}$.

This is the value of $\frac{x}{z}$, and as $\frac{x}{z}$ must be less than unity, because AB is greater than AD, the negative value of u, or $-\frac{1}{2}-\sqrt{\frac{11}{12}}$, is excluded; so that $u=-\frac{1}{2}+\sqrt{\frac{11}{12}}=.45761$ nearly.

Now $u = \frac{AD}{AB}$ = the cosine of the angle BAD, or half the angle of the cone; therefore that angle $= 62° \, 46'$ nearly.

As the tangent of 62° 46′ is not far from being double of the radius, therefore the cone of greatest attraction has the radius of its base nearly double of its altitude.

To compare the attraction of this cone with that of a sphere containing the same quantity of matter, we must express the attraction in terms of u, the ratio of x to z, which has now been found.

Because $\pi x(z^2 - x^2) = 3m^3$, and $z = \frac{x}{u}$, $\pi x\left(\frac{x^2}{u^2} - x^2\right) = \pi x^3\left(\frac{1-u^2}{u^2}\right) = 3m^3$, and $x = m \cdot \sqrt[3]{\frac{3u^2}{\pi(1-u^2)}}$.

Now, we have $A = 2\pi\left(x - \frac{x^2}{z}\right)$, and since $\frac{x}{z} = u$, $\frac{x^2}{z} = mu\sqrt[3]{\frac{3u^2}{\pi(1-u^2)}}$, and $A = 2\pi\left(m \cdot \sqrt[3]{\frac{3u^2}{\pi(1-u^2)}} - mu \cdot \sqrt[3]{\frac{3u^2}{\pi(1-u^2)}}\right) = 2\pi m \cdot (1-u)\sqrt[3]{\frac{3u^2}{\pi(1-u^2)}}$; wherefore, $A^3 = 8\pi^3 m^3(1-u)^3 \times \frac{3u^2}{\pi(1-u^2)} = 24\pi^2 m^3 \cdot \frac{u^2(1-u)^2}{1+u}$.

But if A′ be the attraction of a sphere of which the mass is m^3, on a particle at its surface, $A' = m\sqrt[3]{\frac{16\pi^2}{9}}$, and $A'^3 = m^3 \cdot \frac{16\pi^2}{9}$. Therefore $A^3 : A'^3 :: \frac{24u^2(1-u)^2}{1+u} : \frac{16}{9} :: \frac{27u^2(1-u)^2}{2(1+u)} : 1$; and consequently $A : A' :: 3\sqrt[3]{\frac{u^2(1-u)^2}{2(1+u)}} : 1$.

If, in this expression, we substitute .45761 for u, we shall have $A : A' :: .82941 : 1$, so that the attraction of the cone, when a maximum is about $\frac{4}{5}$ of the attraction of a sphere of equal solidity.

13. Of all the cylinders given in mass, or quantity of matter, to find that which shall attract a particle, at the extremity of its axis, with the greatest force.

Let DF (Fig. 26,) be a cylinder of which the axis is AB, if AC be drawn, the attraction of the cylinder on the particle A is $2\pi(AB+BC-AC)$,* and we have therefore to find when $AB+BC-AC$ is a maximum, supposing $AB.BC^2$ to be equal to a given solid.

Let $AB=x$, $BC=y$, then $AC=\sqrt{x^2+y^2}$, and the quantity that is to be a maximum is $x+y-\sqrt{x^2+y^2}$.

We have therefore $\dot{x}+\dot{y}-\frac{x\dot{x}+y\dot{y}}{\sqrt{x^2+y^2}}=0$, and $(\dot{x}+\dot{y})$ $(x^2+y^2)^{\frac{1}{2}}=x\dot{x}+y\dot{y}$, or $\left(1+\frac{\dot{y}}{\dot{x}}\right)(x^2+y^2)^{\frac{1}{2}}=x+y\frac{\dot{y}}{\dot{x}}$.

But since $\pi xy^2=m^3$, $2xy\dot{y}+y^2\dot{x}=0$, or $2x\dot{y}=-y\dot{x}$, and $\frac{\dot{y}}{\dot{x}}=-\frac{y}{2x}$.

* Princip. Lib. I. Prop. 91. Also Simpson's Fluxions, Vol. II. § 379. In the former, the constant multiplier 2π is omitted, as it is in some other of the theorems relating to the attraction of bodies. This requires to be particularly attended to, when these propositions are to be employed for comparing the attraction of solids of different species.

Therefore $\left(1-\frac{y}{2x}\right)(x^2+y^2)^{\frac{1}{2}}=x-y\frac{y}{2x}$, or $(2x-y)$ $(x^2+y^2)^{\frac{1}{2}}=2x^2-y^2$.

As this equation is homogeneous, if we make $\frac{y}{x}=u$, or $y=ux$, both x and y may be exterminated. For we have by substituting ux for y, $(2x-ux)(x^2+u^2x^2)^{\frac{1}{2}}=2x^2-u^2x^2$, or $(2x^2-ux^2)(1+u^2)^{\frac{1}{2}}=2x^2-u^2x^2$, and dividing by x^2, $(2-u)(1+u^2)^{\frac{1}{2}}=2-u^2$; whence squaring both sides, $(4-4u+u^2)(1+u^2)=4-4u^2+u^4$.

From this, by multiplying and reducing, we get $4u^2-9u=-4$, or $u^2-\frac{9}{4}u=-1$; and $u=\frac{9\pm\sqrt{17}}{8}$.

2. The two values of u in this formula create an ambiguity which cannot be removed without some farther investigation. If A be the attraction of the cylinder, then $A=2\pi(x+y-\sqrt{x^2+y^2})$ into which expression, if we introduce u, and exterminate both x and y, by help of the equations $\pi xy^2=m^3$, and $\frac{y}{x}=u$, we get $A=2\pi^{\frac{2}{3}}m\frac{1+u-\sqrt{1+u^2}}{u^{\frac{2}{3}}}$.

Notwithstanding the radical sign in this formula, there is but one value of A, corresponding to each value of u, as the positive root of $\sqrt{1-u^2}$ is not applicable to the physical problem. This is evident, because the attraction must vanish both when $y=0$, and when $x=0$; that is, both when u is nothing, and when it is infinite. This can only happen when $\sqrt{1+u^2}$ is negative.

Farther, the value of A is always positive, (as it ought to be,) $1+u$ being greater than $\sqrt{1+u^2}$, because it is the square-root of $1+2u+u^2$.

3. We may conceive the relation between A and u most clearly, by supposing A to be the ordinate of a curve in which the abscissæ are represented by the successive values of u. Thus, if OP (fig. 27,) $=u$, and PM=A, the locus of M is a curve of the figure OMM′, intersecting the axis at O, and approaching continually to the line of the abscissæ, OR, as an asymptote, when OP or u increases beyond a certain magnitude. This curve will have both a point, as M where PM the ordinate is a maximum, and a point M′ of contrary flexure. At both of these the fluxion of the ordinate is equal to nothing, and this is the reason of the two values of A that have just been found. For as u increases from nothing, A or PM also increases from nothing till it become a maximum, which happens when $u=\frac{9-\sqrt{17}}{8}$. As u continues to increase, A diminishes; when $u=\frac{9+\sqrt{17}}{8}$, the curvature changes its direction, and as u increases from thence to infinity, A diminishes continually.

The attraction is a maximum, therefore, when $u=\frac{9-\sqrt{17}}{8}$, that is, when y is to x, or the radius of the base of the cylinder, to its altitude, as $9-\sqrt{17}$

to 8, or as 5 to 8 nearly. Therefore also the diameter of the base is to the altitude, when the attraction of the cylinder is greatest, as $9-\sqrt{17}$ to 4, or as 5 to 4 nearly.

4. It may be observed, that the curve OMM′, by which we have expounded the attraction of the cylinder, is but a branch of the curve, which is the complete locus of the equation $2\pi^{\frac{2}{3}}m\left(\frac{1+u-\sqrt{1+u^2}}{u^{\frac{2}{3}}}\right)$ $=\text{A}$. The other branch corresponds to $1+u+\sqrt{1+u^2}$, and has the ordinate infinite, both when $u=0$, and when $u=$ infinity. The curve has also two other branches corresponding to the negative values of u.

5. The attraction of the cylinder, when a maximum is now to be compared with that of a sphere of equal solid content.

First, to compute the quantity $\frac{1+u-\sqrt{1+u^2}}{u^{\frac{2}{3}}}$, when $u=\frac{9-\sqrt{17}}{8}=.6096$, since $u^2=.37161$, $1+u^2=$ 1.37161, and $\sqrt{1+u^2}=1.17116$; so that $1+u-\sqrt{1+u^2}=$ $.43844$.

Also, because $u^2=.37161$ $u^{\frac{2}{3}}=.718945$; and therefore $\frac{1+u-\sqrt{1+u^2}}{u^{\frac{2}{3}}}=\frac{4384}{7189}$. Therefore $\text{A}=$

$$(2\pi^{\frac{2}{3}}m)\left(\frac{1+u-\sqrt{1+u^2}}{u^{\frac{2}{3}}}\right)=2\pi^{\frac{2}{3}}m\times\frac{4384}{7189}.$$

Now, if A′ be the attraction of a sphere of the

solidity m, $\text{A}'=\pi^{\frac{2}{3}}m\left(\frac{16}{9}\right)^{\frac{1}{3}}$, and $\text{A}:\text{A}'::\frac{2\times 4.384}{7189}:\left(\frac{16}{9}\right)^{\frac{1}{3}}::\frac{8768}{7189}:1.2114$, or as 1218 to 1211.4; so that the attraction of the cylinder, even when its form is most advantageous, does not exceed that of a sphere, of the same solid content, by more than a hundred and eighty-third part.

6. In a note on one of the letters of G. L. Lesage, published by M. Prevost of Geneva,* the following theorem is given concerning the attraction of a cylinder and a sphere: If a cylinder be circumscribed about a sphere, the particle placed in the extremity of the axis of the cylinder, or at the point of contact of the sphere, and the base of the cylinder, is attracted equally by the sphere, and by that portion of the cylinder which has for its altitude two-thirds of the diameter of the sphere, and of which the solidity is therefore just equal to that of the sphere.

We may investigate this theorém, by seeking the altitude of such a part of the circumscribing cylinder as shall have the same attraction with the sphere at the point of contact. If r be the radius of the sphere, the attraction at any point of its surface, is $\frac{4\pi r}{3}$; and if x be the altitude of the cylin-

* Notice de la Vie de G. L. Lesage de Genève, par P. Prevost, p. 391.

der, and the radius of its base r, then its attraction on a particle at the extremity of its axis is $2\pi(x+r-\sqrt{x^2+r^2})$. Since these attractions are supposed equal, $2\pi(x+r-\sqrt{x^2+r^2})=\frac{4\pi r}{3}$ and $x+r-\sqrt{x^2+r^2}=\frac{2r}{3}$, whence $\frac{2rx}{3}=\frac{8r^2}{9}$, and $x=\frac{4r}{3}$.

The altitude of the cylinder is therefore $\frac{4}{3}$ of the radius, or $\frac{2}{3}$ of the diameter of the sphere, which is Lesage's Theorem.

This cylinder is also known to be equal in solidity to the sphere; but its attraction is not greater than that of the latter, because the proportion of its altitude to the diameter of its base is not that which gives the greatest attraction. Its altitude is to the diameter of its base, as $\frac{4}{3}r$ to $2r$, or 4 to 6; in order to have the greatest effect, it must be as 4 to 5 nearly, (§ 3.)

Notwithstanding, therefore, that the form of the one of these cylinders is considerably different from that of the other, their attractions are very nearly equal; the one of them being the same with that of the sphere, and the other greater than it by about the 183d part. On each side of the form which gives the maximum of attraction, there may be great variations of figure, without much change in the attracting force. A similar property belongs to all quantities near their greatest or least state, but seems to hold especially in what regards the attraction of bodies.

14. In considering the attraction of the Mountain Schehallien, in such a manner as to make a due allowance for the heterogeneity of the mass, it became necessary to determine the attraction of a half cylinder, or of any sector of a cylinder, on a point situated in its axis, in a given direction, at right angles to that axis. The solution of this problem is much connected with the experimental inquiries concerning the attraction of mountains, and affords examples of maxima of the kind that form the principal object of this paper. The following lemma is necessary to the solution.

Let the quadrilateral DG (Fig. 28,) be the indefinitely small base of a column DH, which has everywhere the same section, and is perpendicular to its base DG.

Let A be a point at a given distance from D, in the plane DG; it is required to find the force with which the column DH attracts a particle at A, in the direction AD.

Let the distance AD$=r$, the angle DAE$=\varphi$, DE (supposed variable) $=y$, and let EF be a section of the solid, parallel and equal to the base DG; and let the area of DG$=m^2$.

The element of the solid DF is $m^2\dot{y}$; and since

DE, or $y=r\tan\varphi$, $\dot{y}=r\dot{\overline{\tan\varphi}}=r\frac{\dot{\varphi}}{\cos\varphi^2}$, so that the element of the solid $=m^2r.\frac{\dot{\varphi}}{\cos\varphi^2}$.

This quantity divided by AE², that is, since AE : AD : : 1 : $\cos\varphi$, by $\frac{r^2}{\cos\varphi^2}$, gives the element of the attraction in the direction AE equal to $\frac{m^2r\dot{\varphi}}{\cos\varphi^2}\times\frac{\cos\varphi^2}{r^2}=\frac{m^2\dot{\varphi}}{r}$. To reduce this to the direction AD, it must be multiplied into the cosine of the angle DAE or φ, so that the element of the attraction of the column in the direction AD is $\frac{m^2}{r}\dot{\varphi}\cos\varphi$, and the attraction itself$=\frac{m^2}{r}\int\dot{\varphi}\cos\varphi=\frac{m^2}{r}\sin\varphi$.

When φ becomes equal to the whole angle subtended by the column, the total attraction is equal to the area of the base divided by the distance, and multiplied by the sine of the angle of elevation of the column.

If the angle of elevation be 30°, the attraction of the column is just half the attraction it would have, supposing it extended to an infinite height.

In this investigation, m^2 is supposed an infinitesimal; but if it be of a finite magnitude, provided it be small, this theorem will afford a sufficient approximation to the attraction of the column, supposing the distance AD to be measured from the centre of gravity of the base, and the angle φ to be that which is subtended by the axis of the column, or by its perpendicular height above the base.

15. Let the semicircle CBG (Fig. 29,) having the centre A, be the base of a half cylinder standing perpendicular to the horizon, AB a line in the plane of the base, bisecting the semicircle, and representing the direction of the meridian; it is required to find the force with which the cylinder attracts a particle at A, in the direction AB, supposing the radius of the base, and the altitude of the cylinder to be given.

Let DF be an indefinitely small quadrilateral, contained between two arches of circles described from the centre A, and two radii drawn to A; and let a column stand on it of the same height with the half cylinder, of which the base is the semicircle CBG. Let $z=$ the angle BAD, the azimuth of D; $v=$ the vertical angle subtended by the column on DF; $a=$ the height of that column, or of the cylinder, AD $=x$, AB, the radius of the base, $=r$.

By the last proposition, the column standing on DF, exerts on A an attraction in the direction AD, which is $=\frac{\mathrm{D}d \times \mathrm{D}f}{\mathrm{AD}} \times \sin v$.

Now $\mathrm{D}d=\dot{x}$, $\mathrm{D}f=x\dot{z}$, and $\mathrm{D}d \times \mathrm{D}f=x\dot{z}\dot{x}$. Therefore the attraction in the direction AD is $\frac{x\dot{x}\dot{z}}{x} \times \sin v=\dot{x}\dot{z}\sin v$, and reduced to the direction AB, it is $\dot{x}\dot{z}\sin v \times \cos z$.

This is the element of the attraction of the cylindric shell or ring, of which the radius is AD or x, and the thickness $\dot{x}$; and therefore integrated on the supposition that z only is variable, and x and v constant, it gives $\dot{x}\sin v\int \dot{z}\cos z = \dot{x}\sin v \times \sin z$ for the attraction of the shell. When $z=90$, and $\sin z =1$, we have the attraction of a quadrant of the shell $=\dot{x}\sin v$, and therefore that of the whole semicircle $=2\dot{x}\sin v$.

Next, if x be made variable, and consequently v, we have $2\int \dot{x}\sin v$ for the attraction of the semi-cylinder.

Now the angle v would have a for its sine if the radius were $\sqrt{a^2+x^2}$, and so $\sin v = \frac{a}{\sqrt{a^2+x^2}}$; wherefore the above expression is $\int \frac{2a\dot{x}}{\sqrt{a^2+x^2}} = 2a\mathrm{L}\,(x+\sqrt{a^2+x^2})+\mathrm{C}$; and as this must vanish when $x=0$, $2a\mathrm{L}a+\mathrm{C}=0$, and $\mathrm{C}=-2a\mathrm{L}a$, so that the fluent is $2a\mathrm{L}\frac{x+\sqrt{a^2+x^2}}{a}$, which, when $x=r$, gives the attraction of the semi-cylinder $=2a\mathrm{L}\frac{r+\sqrt{a^2+r^2}}{a}$.

This expression is very simple, and very convenient in calculation. It is probably needless to remark, that the logarithms meant are the hyperbolic.

16. Let it be required to find the figure of a

solidity which shall attract a particle at its pole with the greatest force.

Let there be an oblate spheroid generated by the revolution of the ellipsis ADBE, (fig. 30,) about the conjugate axis AB, and let F be the focus; then if AF be drawn, and the arch CG described from the centre A, the force with which the spheroid draws a particle at A, in the direction AC, is $\frac{4\pi . AC . CD^2}{CF^3}(CF - CG^*)$. (Maclaurin's Fluxions, § 650.)

Let this force $= F$, $AC = a$, $CD = b$, the angle CAF $= \varphi$; then $CF = a \tan\varphi$, and $F = \frac{4\pi ab^2}{a^3 \tan\varphi^3}(\tan\varphi - \varphi)a = \frac{4\pi b^2}{a} \cdot \frac{\tan\varphi - \varphi}{\tan\varphi^3}$.

Now if m^3 be the solidity of the spheroid, since that solidity is two-thirds of the cylinder, having CD for the radius of its base, and AB for its altitude; therefore $m^3 = \frac{2}{3} \times \pi b^2 \times 2a = \frac{4}{3}\pi ab^2$; so that $b^2 = \frac{m^3}{\frac{4}{3}\pi a} = \frac{3m^3}{4\pi a}$, and $\frac{b^2}{a} = \frac{3m^3}{4\pi a^2}$.

But because $AF : AC :: 1 : \cos\varphi$, or $b : a :: 1 : \cos\varphi$, $b^2 = \frac{a^2}{\cos\varphi^2}$, and $\frac{b^2}{a} = \frac{a}{\cos\varphi^2}$.

Now since $b^2 = \frac{a^2}{\cos\varphi^2}$, and also $b^2 = \frac{3m^3}{4\pi a}$, we have

---

* The multiplier 2π, omitted by Maclaurin, is restored as above, § 13.

$\frac{a^2}{\cos\varphi^2}=\frac{3m^3}{4\pi a}$, and $a^3=\frac{3m^3}{4\pi}\cos\varphi^2$, or if $\frac{3m^3}{4\pi}=n^3$, $a^3=n^3\cos\varphi^2$, and $a=n\cos\varphi^{\frac{2}{3}}$.

Hence, as $\frac{b^2}{a}=\frac{a}{\cos\varphi^2}$, $\frac{b^2}{a}=\frac{n\cos\varphi^{\frac{2}{3}}}{\cos\varphi^2}=\frac{n}{\cos\varphi^{\frac{4}{3}}}$.

By substituting this value of $\frac{b^2}{a}$ in the value of F, we have $F=\frac{4\pi n}{\cos\varphi^{\frac{4}{3}}}\times\frac{\tan\varphi-\varphi}{\tan\varphi^3}$, and because $\tan\varphi^3=\frac{\sin\varphi^3}{\cos\varphi^3}$, $F=\frac{2\pi n(\tan\varphi-\varphi)\cos\varphi^3}{\sin\varphi^3\cos\varphi^{\frac{4}{3}}}=\frac{2\pi n(\tan\varphi-\varphi)\cos\varphi^{\frac{5}{3}}}{\sin\varphi^3}=2\pi n(\tan\varphi-\varphi)\cos\varphi^{\frac{5}{3}}\sin\varphi^{-3}$.

Now, when the product of any number of factors is a maximum, if the fluxion of each factor be divided by the factor itself, the sum of the quotients is equal to nothing. Therefore

$$\frac{\frac{\dot\varphi}{\cos\varphi^2}-\dot\varphi}{\tan\varphi-\varphi}+\frac{5\cos\varphi^{\frac{2}{3}}\dot{\overline{\cos\varphi}}}{3\cos\varphi^{\frac{5}{3}}}-\frac{3\sin\varphi^{-4}\dot{\overline{\sin\varphi}}}{\sin\varphi^{-3}}=0, \text{ or}$$

$\frac{\dot\varphi(1-\cos\varphi^2)}{\cos\varphi^2(\tan\varphi-\varphi)}-\frac{5\dot\varphi\sin\varphi}{3\cos\varphi}-\frac{3\dot\varphi\cos\varphi}{\sin\varphi}=0$, and $\frac{\sin\varphi^2}{\cos\varphi^2(\tan\varphi-\varphi)}-\frac{5\sin\varphi}{3\cos\varphi}-\frac{3\cos\varphi}{\sin\varphi}=0$; $\frac{\sin\varphi}{\cos\varphi(\tan\varphi-\varphi)}=\frac{5}{3}+3\frac{\cos\varphi^2}{\sin\varphi^2}$, and $\frac{\tan\varphi}{\tan\varphi-\varphi}=\frac{5}{3}+3\cot\varphi^2$.

Hence $\frac{3\tan\varphi}{5+9\cot\varphi^2}=\tan\varphi-\varphi$, and $\varphi=\tan\varphi-\frac{3\tan\varphi}{5+9\cot\varphi^2}=\frac{5\tan\varphi+9\tan\varphi\cot\varphi^2-3\tan\varphi}{5+9\cot\varphi^2}=\frac{2\tan\varphi+9\cot\varphi}{5+9\cot\varphi^2}$.

Let $\tan\varphi = t$, then $\varphi = \frac{2t+\frac{9}{t}}{5+\frac{9}{t^2}} = \frac{2t^3+9t}{5t^2+9} = \frac{t(9+2t^2)}{9+5t^2}$;

which, therefore, is the value of φ when F is a maximum.

The value of φ, now found, is remarkable for being a near approximation to any arch of which t is the tangent, provided that arch do not exceed 45°. The less the arch is, the more near is the approximation; but the expression can only be considered as accurate when $\varphi=0$.

This will be madé evident by comparing the fraction $\frac{t(9+2t^2)}{9+5t^2}$ with the series, that gives the arch in terms of the tangent t, viz. $\varphi = t - \frac{t^3}{3} + \frac{t^5}{5} - \frac{t^7}{7} +$, &c.

The fraction $\frac{t(9+2t^2)}{9+5t^2} = t - \frac{t^3}{3} + \frac{5t^5}{27} - \frac{5^2t^7}{3.9^2} +$, &c.

The two first terms of these series agree; and in the third terms, the difference is inconsiderable, while t is less than unity; but the agreement is never entire, unless $t=0$, when both series vanish.

The attraction, therefore, or the gravitation at the pole of an oblate spheroid, is not a maximum, until the eccentricity of the generating ellipsis vanish, and the spheroid pass into a sphere.

From the circumstance of the value of φ above found, agreeing nearly with an indefinite number of arches, we must conclude, that when a sphere passes into an oblate spheroid, its attraction varies at first exceeding slowly, and continues to do so till its oblateness, or the eccentricity of the generating ellipsis, become very great. This may be shown, by taking the value of F, and substituting in it that of φ, in terms of $\tan\varphi$.

We have $F=\frac{4\pi n}{\cos\varphi^{\frac{4}{3}}}\times\frac{\tan\varphi-\varphi}{\tan\varphi^3}$; and since $\varphi=\tan\varphi-\frac{\tan\varphi^3}{3}+\frac{\tan\varphi^5}{5}-$ &c., $\tan\varphi-\varphi=\frac{\tan\varphi^3}{3}-\frac{\tan\varphi^5}{5}+$ &c., and $F=\frac{4\pi n}{\cos\varphi^{\frac{4}{3}}}\left(\frac{\tan\varphi^3}{3}-\frac{\tan\varphi^5}{5}+\frac{\tan\varphi^7}{7}\right)\frac{1}{\tan\varphi^3}=\frac{4\pi n}{\cos\varphi^{\frac{4}{3}}}$ $\left(\frac{1}{3}-\frac{\tan\varphi^2}{5}+\frac{\tan\varphi^4}{7}\right)$. When $\varphi=0$, $F=\frac{4\pi n}{3}$; and since $n=m\sqrt[3]{\frac{3}{4\pi}}$, $F=\frac{4\pi m}{3}\sqrt[3]{\frac{3}{4\pi}}=m\sqrt[3]{\frac{64\pi^3\times 3}{27\times 4\pi}}=$ $m\sqrt[3]{\frac{16\pi^2}{9}}$, which is the attraction at the surface of a sphere of the solidity m^3, as was already shown. This last is the conclusion we had to expect, the spheroid, when it ceases to have any oblateness, becoming of necessity a sphere.

It is evident also, that the variations of φ will but little affect the magnitude of F, while φ and $\tan\varphi$ are small, as the least power of $\tan\varphi$ that enters into the value of F is the square.

For, instead of $\cos\varphi^{-\frac{4}{3}}$, we may, when φ is very small, write $1+\frac{2}{3}\tan\varphi^2$; so that $F=4\pi n\left(1+\frac{2}{3}\tan\varphi^2\right)$ $\left(\frac{1}{3}-\frac{\tan\varphi^2}{5}+\frac{\tan\varphi^4}{7}-\text{ \&c. }\right)$

If the oblateness of a spheroid diminish, while its quantity of matter remains the same, its attraction will increase till the oblateness vanish, and the spheroid become a sphere, when the attraction at its poles, as we have seen, becomes a maximum. If the polar axis continue to increase, the spheroid becomes oblong, and the attraction at the poles again diminishes. This we may safely conclude from the law of continuity, though the oblong spheroid has not been immediately considered.

18. To find the force with which a particle of matter is attracted by a parallelepiped, in a direction perpendicular to any of its sides.

First, let EM (Fig. 31,) be a parallelepiped, having the thickness CE indefinitely small, A, a particle situated any where without it, and AB a perpendicular to the plane CDMN. The attraction in the direction AB is to be determined.

Let the solid EM be divided into columns perpendicular to the plane NE, having indefinitely small rectangular bases, and let CG be one of those columns.

If the angle CAB, the azimuth of this column relatively to AB, be called z, CAD, its angle of elevation from A, e, and m^2, the area of the little rectangle CF; then, as has been already shown, the attraction of the column CG, in the direction AC, is $\frac{m^2}{\mathrm{AC}}\sin e$; and that same attraction, reduced to the direction AB, is $\frac{\dot{m}^2}{AC}\sin e \cos z$. This is the element of the attraction of the solid, and if we call that attraction $\dot{f}$, $\dot{f}=\frac{\dot{m}^2}{AC}\sin e\cos z$. .

Now, if AB$=a$, because $1:\cos z::AC:AB$, $AC=\frac{a}{\cos z}$; so that $\dot{f}=\frac{\dot{m}^2}{a}\sin e\cos z^2$.

But BC$=a\tan z$; and therefore KC, the fluxion of BC, is $=a\frac{\dot{z}}{\cos z^2}$; if, then, $CE=n$, $\dot{m}^2=CE\times CK=na\frac{\dot{z}}{\cos z^2}$, and substituting this for $\dot{m}^2$, we get $\dot{f}=n\dot{z}\sin e$.

Next, to express sin$e$, in terms of z, if we make E=BAL, the angle subtended by the vertical columns, when it is greatest, or the inclination of the plane ADM, to the plane ACN, then we may consider the angle CAD, as measured by the side of a right angled spherical triangle, of which the other side is 90—z, and E the angle, adjacent to that side, and therefore $\tan e=\sin(90-z)\tan E=\cos z\tan E$.

But $\tan E=\tan BAL=\frac{BL}{BA}=\frac{b}{a}$, supposing BL, or DC$=b$.

Therefore $\tan e=\frac{b}{a}\cos z$, or $\frac{\sin e}{\cos e}=\frac{b}{a}\cos z$.

Hence $\frac{\sin e^2}{\cos e^2}=\frac{b^2}{a^2}\cos z^2$, or $\frac{\sin e^2}{1-\sin e^2}=\frac{b^2}{a^2}\cos z^2$, and

$\text{sine}^2 = \frac{b^2}{a^2}\cos z^2 - \text{sine}^2\frac{b^2}{a^2}\cos z^2$; and therefore $\text{sine} =$

$$\frac{\frac{b}{a}\cos z}{\sqrt{1+\frac{b^2}{a^2}\cos z^2}}.$$

If this value of sine be substituted for it, we have $\dot{f} = n\dot{z}\,\text{sine} = \frac{bn\dot{z}\cos z}{a\sqrt{1+\frac{b^2}{a^2}\cos z^2}}.$

Let $u = \sin z$, then $\dot{u} = \dot{z}\cos z$, and $\cos z^2 = 1 - u^2$; wherefore, again, by substitution, $\dot{f} = \frac{bn\dot{u}}{a\sqrt{1+\frac{b^2}{a^2}(1-u^2)}} =$

$\frac{bn\dot{u}}{\sqrt{a^2+b^2-b^2u^2}}$. Let a^2+b^2, or $AL^2 = c^2$, then $\dot{f} =$

$$\frac{bn\dot{u}}{\sqrt{c^2-b^2u^2}} = \frac{bn\dot{u}}{c\sqrt{1-\frac{b^2}{c^2}u^2}}.$$

If, therefore, φ be such an arch, that $\frac{bu}{c} = \sin\varphi$, $\frac{b\dot{u}}{c} = \dot{\varphi}\cos\varphi$, and $\sqrt{1-\frac{b^2}{c^2}u^2} = \cos\varphi$, then $\frac{bn\dot{u}}{c\sqrt{1-\frac{b^2}{c^2}u^2}} =$

$\frac{n\dot{\varphi}\cos\varphi}{\cos\varphi} = n\dot{\varphi}$. Thus, $\dot{f} = n\dot{\varphi}$, and $f = n\varphi + B$, B being a constant quantity.

Now, since $\sin\varphi = \frac{bu}{c} = \frac{b}{c}\sin z$, φ is nothing when z is nothing; and as f may be supposed to begin when z begins, we have likewise $B=0$; and $f = n\varphi = n$ multiplied into an arch, the sine of which is to the sine of z, in the given ratio of c to b. Or f is such that $\sin\frac{f}{n}=\frac{b}{c}\times\frac{BC}{AC}=\frac{BL}{AL}\times\frac{BC}{AC}$.

Hence this rule, multiply the sine of the greatest elevation, into the sine of the greatest azimuth of the solid; the arch of which this is the sine, multiplied into the thickness of the solid, is equal to its attraction in the direction of the perpendicular from the point attracted.

The heighth and the length of the parallelepiped, are, therefore, similarly involved in the expression of the force, as they ought evidently to be from the nature of the thing.

19. This theorem leads directly to the determination of the attraction of a pyramid, having a rectangular base, on a particle at its vertex. For if we consider EM (Fig. 31,) as a slice of a pyramid parallel to its base, A being the vertex, then the slice behind EM subtending the same angles that it does, will have its force of attraction $=n'\varphi$, n' being its thickness, and so of all the rest; and, therefore, the sum of all these attractions, if p denote the whole height of the solid, or the perpendicular from A on its base, will be $p\varphi$. But as $n\varphi$ is only the attraction of the part HB, it must be doubled to give the attraction of the whole solid EM, which is, therefore, $2n\varphi$; and this must again be doubled, to give the attraction of the part which is on the side of AB, opposite to EM; thus the element of the

attraction of the pyramid is $4n\varphi$, and the whole attraction corresponding to the depth p, is $4p\varphi$.

If the solid is the frustum of a pyramid whose depth is p', and vertex A, the angle φ being determined as before, the attraction on A is $4p'\varphi$.

If we suppose BC and BL to be equal, and therefore the angle BAL= the angle BAC, calling either of them η, then $\sin\varphi=\sin\eta^2$, by what has been already shown; and from this equation, as η is supposed to be given, φ is determined.

This expression for the attraction of an isosceles pyramid, having a rectangular base, may be of use in many computations concerning the attraction of bodies.

If the solidity of the pyramid be given, from the equations $f=4p\varphi$, and $\sin\varphi=\sin\eta^2$, we may determine η, and p, that is, the form of the pyramid when f is a maximum.

Let the solidity of the pyramid $=m^3$, then p, being the altitude of the pyramid, and η half the angle at the vertex, $p\tan\eta=$ half the side of the base, (which is a square,) and therefore the area of the base $=4p^2\tan\eta^2$, and the solidity of the pyramid $\frac{4}{3}p^3\tan\eta^2$; so that $\frac{4}{3}p^3\tan\eta^2=m^3$.

Now $\tan\eta^2=\frac{\sin\eta^2}{\cos\eta^2}$, and $\sin\varphi=\sin\eta^2$, also $1-\sin\varphi=1-\sin\eta^2=\cos\eta^2$, therefore $\tan\eta^2=\frac{\sin\varphi}{1-\sin\varphi}$; so that $m^3=\frac{4}{3}p^3$

$\left(\frac{\sin\varphi}{1-\sin\varphi}\right)$, and $p^3=\frac{3}{4}m^3\left(\frac{1-\sin\varphi}{\sin\varphi}\right)$ or $p=m\sqrt[3]{\frac{3(1-\sin\varphi)}{4\sin\varphi}}$; we have, therefore, f, that is $4p\varphi=4m\varphi\sqrt[3]{\frac{3(1-\sin\varphi)}{4\ \sin\varphi}}$. This last is, therefore, a maximum by hypothesis; and, consequently, its cube, or $64m^3\varphi^3\times\frac{3(1-\sin\varphi)}{4\sin\varphi}$, or omitting the constant multipliers, $\varphi^3\left(\frac{1-\sin\varphi}{\sin\varphi}\right)$ must be a maximum.

If we take the fluxion of each of these multipliers, and divide it by the multiplier itself, and put the sum equal to nothing, we shall have $\frac{3\dot\varphi}{\varphi}-\frac{\dot\varphi\cos\varphi}{1-\sin\varphi}-\frac{\dot\varphi\cos\varphi}{\sin\varphi}=0$, or $\frac{3}{\varphi}=\frac{\cos\varphi}{1-\sin\varphi}+\frac{\cos\varphi}{\sin\varphi}=\frac{\cos\varphi\sin\varphi+\cos\varphi-\cos\varphi\sin\varphi}{\sin\varphi(1-\sin\varphi)}=\frac{\cos\varphi}{\sin\varphi(1-\sin\varphi)}$, and inverting these fractions $\frac{\varphi}{3}=\frac{\sin\varphi(1-\sin\varphi)}{\cos\varphi}=\tan\varphi\,(1-\sin\varphi)$, or $\varphi=3\tan\varphi(1-\sin\varphi)$.

The solution of this transcendental equation may easily be obtained, by approximation, from the trigonometric tables, if we consider that $1-\sin\varphi$ is the coversed sine of φ. Thus taking the logarithms, we have $L\varphi=L3+L\tan\varphi+L\text{covers}\,\varphi$. From which, by trial, it will soon be discovered, that φ is nearly equal to an arch of 48°. To obtain a more exact value of φ, let $\varphi=\text{arc}(48°+\beta)$, β being a number of minutes to be determined. Because $\text{arc}48°=$

.8377580, and $\text{arc}(48^\circ+\beta)=.8377580+.0002909\beta$, therefore $\log.\text{arc}(48^\circ+\beta)=9.9231186+.0001506\beta$.

In the same manner,

$$\text{Ltan}(48^\circ+\beta)=0.0455626+.0002540\beta,$$

$$\text{and Lcovers}(48^\circ+\beta)=9.4096883-.0003292\beta$$

$$\text{L}3=0.4771213$$

$$\text{Sum}=9.9323722-.0000752\beta$$

$$\text{Subtract } \log.\text{arc}(48^\circ+\beta)=9.9231186+.0001506\beta$$

$$\text{Remainder} = .0092536-.0002258\beta=0.$$

Whence $\beta=\frac{92536}{2258}=41'$ nearly.

A second approximation will give a correction $=-20''$, so that $\varphi=\text{arc}.48^\circ40'\frac{2}{3}$; and since $\sin\varphi=\sin\eta^2$, $\sin\eta=\sqrt{\sin\varphi}$, so that $\eta=76^\circ.30'$, and 2η, or the whole angle of the pyramid $=153^\circ$.

An isosceles pyramid, therefore, with a square base, will attract a particle at its vertex with greatest force, when the inclination of the opposite planes to one another is an angle of 153°.

20. To return to the attraction of the parallelepiped, it may be remarked, that the theorem concerning this attraction already investigated, § 18, though it applies only to the case when the parallelepiped is indefinitely thin, leads, nevertheless, to some very general conclusions. It was shown, that the attraction which the solid EL (Fig. 31) exerts on the particle A, in the direction AB, is $n\varphi$, φ being an arch, such that $\sin\varphi=\sin \text{BAC}\times\sin \text{BAL}=\sin z \sin \text{E}$; and, therefore, if B be the centre of a rect-

angle, of which the breadth is 2BC, and the height 2BL, the attraction of that plane, or of the thin solid, having that plane for its base, and n, for its thickness, is $4n\varphi$. Now, φ, which is thus proportional to the attraction of the plane, is also proportional to the spherical surface, or the angular space, subtended by the plane at the centre A.

For suppose PSQ (fig. 32,) and OQ to be two quadrants of great circles of a sphere, cutting one another at right angles in Q; let QS$=$E, and QR$=z$. Through S, and O the pole of PSQ, draw the great circle OST, and through P and R, the great circle PTR, intersecting OS in T. The spherical quadrilateral SQRT, is that which the rectangle CL (fig. 31) would subtend, if the sphere had its centre at A, if the point Q was in the line AB, and the circle PQ, in the vertical plane ABL.

Now, in the spherical triangle PST, right angled at S, $\cos T = \cos PS \sin SPT = \sin QS \sin QR = \sin E \sin z$. But this is also the value of $\sin\varphi$, and therefore φ is the complement of the angle T, or $\varphi = 90 - T$.

But the area of the triangle PQR, in which both Q and R are right angles, is equal to the rectangle under the arch QR, which measures the angle QPR, and the radius of the sphere. Also the area SPT$=$arc$(S+T+P-180°)r$; that is, because S is a right angle, $=$arc$(T+P-90)r=$

arc $(T+QR-90)r$; and taking this away from the triangle PQR, there remains the area QSTR= arc $(QR-T-QR+90°)r=(90°-T)r=\varphi r$. The arch φ, therefore, multiplied into the radius, is equal to the spherical quadrilateral QSTR, subtended by the rectangle BD.

This proposition is evidently applicable to all rectangles whatsoever For when the point B, where the perpendicular from A meets the plane of the rectangle, falls anywhere, as in fig. 35, then it may be shown of each of the four rectangles BD, BM, BM′, BD′, which make up the whole rectangle DM′, that its attraction in the direction AB is expounded by the area of the spherical quadrilateral subtended by it, and, therefore, that the attraction of the whole rectangle MD′, is expounded by the sum of these spherical quadrilaterals, that is, by the whole quadrilateral subtended by MD′. In the same manner, if the perpendicular from the attracted particle, were to meet the plane without the rectangle MD′, the difference between the spherical quadrilaterals subtended by MC and M′C, would give the quadrilateral, subtended by the rectangle MD′, for the value of the attraction of that rectangle.

Therefore, in general, *if a particle* A, *gravitate to a rectangular plane, or to a solid indefinitely thin, contained between two parallel rectangular planes, its gravitation, in the line per-*

pendicular to those planes, will be equal to the thickness of the solid, multiplied into the area of the spherical quadrilateral subtended by either of those planes at the centre A.

The same may be extended to all planes, by whatever figure they be bounded, as they may all be resolved into rectangles of indefinitely small breadth, and having their lengths parallel to a straight line given in position.

The gravitation of a point toward any plane, in a line perpendicular to it, is, therefore, equal to n, a quantity that expresses the intensity of the attraction, multiplied into the area of the spherical figure, or, as it may be called, the angular space subtended by the given plane.

Thus, in the case of a triangular plane, where the angles subtended at A, by the sides of the triangle, are a, b, and c; since Euler has demonstrated* that the area of the spherical triangle contained by these arches, is equal to the rectangle under the radius, and an arch Δ, such that $\cos\frac{1}{2}\Delta = \frac{1+\cos a+\cos b+\cos c}{4\cos\frac{1}{2}a\cos\frac{1}{2}b\cos\frac{1}{2}c}$; if Δ be computed, the attraction $= n\Delta$.

In the case of a circular plane, our general proposition agrees with what Sir Isaac Newton has demonstrated. If CFD (fig. 33,) be a circle, BA

* Nov. Acta Petrop. 1792, p. 47.

a line perpendicular to the plane of it from its centre B; A, a particle anywhere in that line; the force with which A is attracted, in the direction AB, is $2\pi\left(1-\frac{AB}{AD}\right)$,* in which the multiplier 2π is supplied, being left out in the investigation referred to, where a quantity only proportional to the attraction is required. Now $\frac{AB}{AD}$ is the cosine of the angle BAD, and, therefore, $1-\frac{AB}{AD}$ is its versed sine; and, therefore, if the arch GEK be described from the centre A, with the radius 1, and if the sine GH, and the chord EG be drawn, HE is the versed sine of BAD, and the attraction $=2\pi EH$. But $2EH=EG^2$, because 2 is the diameter of the circle GEK; therefore the attraction $=\pi EG^2=$ the area of the circle of which EG is the radius, or the spherical surface included by the cone, which has A for its vertex, and the circle CFD for its base.

21. From the general proposition, that the attraction of any plane figure, whatever its boundary may be, in a line perpendicular to the plane, is at any distance proportional to the angular space, or to the area of the spherical figure which the plane figure subtends at that distance, we can easily de-

* Princip. Lib. i. Prop. 90.

duce a demonstration of this other proposition, that whatever be the figure of any body, its attraction will decrease in a ratio that approaches continually nearer to the inverse ratio of the squares of the distances, as the distances themselves are greater. In other words, the inverse ratio of the squares of the distances, is the limit to which the law by which the attraction decreases, continually approaches as the distances increase, and with which it may be said to coincide when the distances are infinitely great.

This proposition, which we usually take for granted, without any other proof, I believe, than, some indistinct perception of what is required by the law of continuity, may be rigorously demonstrated from the principle just established.

Let B (fig. 34,) be a body of any figure whatsoever, A a particle situated at a distance from B vastly greater than any of the dimensions of B, so that B may subtend a very small angle at A; from C, a point in the interior of the body, suppose its centre of gravity, let a straight line be drawn to A, and let A′ be another point, more remote from B than A is, where a particle of matter is also placed.

The directions in which A and A′ gravitate to B, as they must tend to some point within B, must either coincide with AC, or make a very small angle with it, which will be always the less, the greater the distance.

Let the body B be cut by two planes, at right angles to AC, and indefinitely near to one another, so as to contain between them a slice or thin section of the body, to which A and A′ may be considered as gravitating, nearly in the direction of the line AC perpendicular to that section.

The gravitation of A, therefore, to the aforesaid section, will be to that of A′ to the same, as the angular space subtended by that section at A, to the angular space subtended by it at A′. But these angular spaces, when the distances are great, are inversely as the squares of those distances, and therefore, also, the gravitation of A toward the section, will be to that of A′ inversely as the squares of the distances of A and A′ from the section. Now these distances may be accounted equal to CA and CA′, from which they can differ very little, wherever the section is made.

The gravitations of A and A′ toward the said section, are, therefore, as $\frac{1}{AC^2}$ to $\frac{1}{A'C^2}$. And the same may be proved of the gravitation to all the other sections, or laminæ, into which the body can be divided by planes perpendicular to AC; therefore the sums of all these gravitations, that is, the whole gravitations of A to B, and of A′ to B, will be in that same ratio, that is, as $\frac{1}{AC^2}$ to $\frac{1}{A'C^2}$, or inversely as the squares of the distances from C. Q. E. D.

It is evident that the greater the distances AC, A′C are, the nearer is this proposition to the truth, as the quantities rejected in the demonstration, become less in respect of the rest, in the same proportion that AC and A′C increase.

It is here assumed, that the angular space subtended by the same plane figure, is inversely as the square of the distance. This proposition may be proved to be rigorously true, if we consider the inverse ratio of the squares of the distances, as a limit to which the other ratio constantly converges.

It is a proposition also usually laid down in optics, where the *visible space* subtended by a surface, is the same with what we have here called the *angular space* subtended by it, or the portion of a spherical superficies that would be cut off by a line passing through the centre of the sphere, and revolving round the boundary of the figure. The centre of the sphere is supposed to coincide with the eye of the observer, or with the place of the particle attracted, and its radius is supposed to be unity.

The propositions that have been just now demonstrated concerning the attraction of a thin plate contained between parallel planes, have an immediate application to such inquiries concerning the attraction of bodies, as were lately made by Mr Cavendish.

In some of the experiments instituted by that ingenious and profound philosopher, it became necessary to determine the attraction of the sides of a wooden case, of the form of a parallelepiped, on a body placed anywhere within it. (Philosophical Transactions, 1798, p. 523.) The attraction in the direction perpendicular to the side, was what occasioned the greatest difficulty, and Mr Cavendish had recourse to two infinite series, in order to determine the quantity of that attraction. The determination of it, from the preceding theorems, is easier and more accurate.

Let MD′ (Fig. 35,) represent a thin rectangular plate, A a particle attracted by it, AB a perpendicular on the plane MD′, NBC, LBL′, two lines drawn through B parallel to the sides of the rectangle MD′. Let AC, AL, AN, AL′ be drawn.

Then, if we find φ such that $\sin\varphi = \frac{BL}{AL} \times \frac{BC}{AC}$, the attraction of the rectangle CL is $n\varphi$, n denoting the thickness of the plate.

So also, if $\sin\varphi' = \frac{BL}{AL} \times \frac{BN}{AN}$, the attraction of LN is $= n\varphi'$.

If $\sin\varphi'' = \frac{BN}{AN} \times \frac{BL'}{AL'}$, the attraction of NL′ is $= n\varphi''$

Lastly, if $\sin\varphi'''=\frac{BL'}{AL'}\times\frac{BC}{AC}$, the attraction of L′C $=n\varphi'''$.

Thus the whole effect of the plane MD′, or $f=n(\varphi+\varphi'+\varphi''+\varphi''')$.

We may either suppose φ, φ', &c. defined as above, or by the following equations, where η, η', η'', &c. denote the angles subtended by the sides of the rectangles that meet in B, beginning with BC, and going round by L, N and L′ to C.

$$\sin\varphi=\sin\eta\,\sin\eta'$$
$$\sin\varphi'=\sin\eta'\,\sin\eta''$$
$$\sin\varphi''=\sin\eta''\,\sin\eta'''$$
$$\sin\varphi'''=\sin\eta'''\sin\eta.$$

If the computation is to be made by the natural sines, it will be better to use the following formulæ:

$$\sin\varphi=\frac{1}{2}\cos(\eta-\eta')-\frac{1}{2}\cos(\eta+\eta')$$

$$\sin\varphi'=\frac{1}{2}\cos(\eta'-\eta'')-\frac{1}{2}\cos(\eta'+\eta'')$$

$$\sin\varphi''=\frac{1}{2}\cos(\eta''-\eta''')-\frac{1}{2}\cos(\eta''+\eta''')$$

$$\sin\varphi'''=\frac{1}{2}\cos(\eta'''-\eta)-\frac{1}{2}\cos(\eta'''+\eta).$$

By either of these methods, the determination of the attraction is reduced to a very simple trigonometrical calculation.

22. The preceding theorems will also serve to determine the attraction of a parallelepiped, of any

given dimensions, in the direction perpendicular to its sides.

Let BF (Fig. 36) be a parallelepiped, and A a point in BK, the intersection of two of its sides, where a particle of matter is supposed to be placed; it is required to find the attraction in the direction AB.

Though the placing of A in one of the intersections of the planes, seems to limit the inquiry, it has in reality no such effect; for wherever A be with respect to the parallelepiped, by drawing from it a perpendicular to the opposite plane of the solid, and making planes to pass through this perpendicular, the whole may be divided into four parallelepipeds, each having AB for an intersection of two of its planes; and being, therefore, related to the given particle, in the same way that the parallelepiped BF is to A.

Let GH be any section of the solid parallel to EC, and let it represent a plate of indefinitely small thickness.

Let $AB'=x$, $B'b$, the thickness of the plate $=\dot{x}$. Then φ being so determined, that $\sin\varphi=\sin B'AH$ $\sin B'AG$, the attraction of the plate GH is $\varphi\dot{x}$, which, therefore, is the element of the attraction of the solid. If that attraction $=F$, then $F=\int\varphi\dot{x}$. But $\int\varphi\dot{x}=\varphi x-\int x\dot{\varphi}$; and the determination of F depends, therefore, on the integration of $x\dot{\varphi}$.

Now $\dot{\varphi}\cos\varphi = \overline{\dot{\sin\varphi}}$, and, therefore, $x\dot{\varphi} = \frac{x\overline{\dot{\sin\varphi}}}{\cos\varphi}$.

If $B'G = b$, and $B'H = \beta$, then $\sin B'AG = \frac{b}{AG} = \frac{b}{\sqrt{b^2+x^2}}$, and $\sin B'AH = \frac{\beta}{AH} = \frac{\beta}{\sqrt{\beta^2+x^2}}$; so that $\sin\varphi = \left(\frac{b}{\sqrt{b^2+x^2}}\right)\left(\frac{\beta}{\sqrt{\beta^2+x^2}}\right)$ and $(\sin\varphi)^2 = \frac{b^2\beta^2}{(b^2+x^2)(\beta^2+x^2)}$.

Hence, $\cos\varphi^2 = 1 - \sin\varphi^2 = 1 - \frac{b^2\beta^2}{(b^2+x^2)(\beta^2+x^2)} = \frac{x^2(b^2+\beta^2+x^2)}{(b^2+x^2)(\beta^2+x^2)}$, and $\cos\varphi = \frac{x\sqrt{b^2+\beta^2+x^2}}{\sqrt{(b^2+x^2)(\beta^2+x^2)}}$.

Again, because $\sin\varphi = \left(\frac{b}{\sqrt{b^2+x^2}}\right)\left(\frac{\beta}{\sqrt{\beta^2+x^2}}\right)$, $\overline{\dot{\sin\varphi}} = \left(\frac{-bx\dot{x}}{(b^2+x^2)^{\frac{3}{2}}}\right)\left(\frac{\beta}{(\beta^2+x^2)^{\frac{1}{2}}}\right) - \left(\frac{\beta x\dot{x}}{(\beta^2+x^2)^{\frac{3}{2}}}\right)\left(\frac{b}{(b^2+x^2)^{\frac{1}{2}}}\right)$.

Hence $\frac{\overline{\dot{\sin\varphi}}}{\cos\varphi}$ or $\dot{\varphi} =$

$$\left(\frac{-b\beta x\dot{x}}{(b^2+x^2)^{\frac{3}{2}}(\beta^2+x^2)^{\frac{1}{2}}} - \frac{b\beta x\dot{x}}{(\beta^2+x^2)^{\frac{3}{2}}(b^2+x^2)^{\frac{1}{2}}}\right)\left(\frac{(b^2+x^2)^{\frac{1}{2}}(\beta^2+x^2)^{\frac{1}{2}}}{x\sqrt{b^2+\beta^2+x^2}}\right) = -\frac{b\beta\dot{x}}{(b^2+x^2)(c^2+x^2)^{\frac{1}{2}}} - \frac{b\beta\dot{x}}{(\beta^2+x^2)(c^2+x^2)^{\frac{1}{2}}},$$

c^2 being put for $b^2+\beta^2$.

Therefore $x\dot{\varphi} = -\frac{b\beta x\dot{x}}{(b^2+x^2)(c^2+x^2)^{\frac{1}{2}}} - \frac{b\beta x\dot{x}}{(\beta^2+x^2)(c^2+x^2)^{\frac{1}{2}}}$.

Now, $\int \frac{-b\beta x\dot{x}}{(b^2+x^2)(c^2+x^2)^{\frac{1}{2}}} = b\log \frac{\beta+\sqrt{c^2+x^2}}{\sqrt{b^2+x^2}} + C$; (Harmonia Mensurarum, Form. IX.); and

$$\int \frac{-b\beta x\dot{x}}{(\beta^2+x^2)(c^2+x^2)^{\frac{1}{2}}} = \beta\log \frac{b+\sqrt{c^2+x^2}}{\sqrt{\beta^2+x^2}} + C.$$

Therefore $\int x\dot{\varphi} = b\log \frac{\beta+\sqrt{c^2+x^2}}{\sqrt{b^2+x^2}} + \beta\log \frac{b+\sqrt{c^2+x^2}}{\sqrt{\beta^2+x^2}} +$ C, and $\int \varphi\dot{x} = \varphi x - b\log \frac{\beta+\sqrt{c^2+x^2}}{\sqrt{b^2+x^2}} - \beta\log \frac{b+\sqrt{c^2+x^2}}{\sqrt{\beta^2+x^2}} - C.$

If, then, we determine C, so that the fluent may begin at K, and end at B; if, also, we make η the value of φ, that corresponds to AB or a; and η', the value of it that corresponds to AK or a', we have the whole attraction of the solid, or

$$F = \eta a - \eta' a' - \left(b\log \frac{\beta+\sqrt{c^2+a^2}}{\sqrt{b^2+a^2}}\right)\left(\frac{\sqrt{b^2+a'^2}}{\beta+\sqrt{c^2+a'^2}}\right)$$
$$-\left(\beta\log \frac{b+\sqrt{c^2+a^2}}{\sqrt{\beta^2+a^2}}\right)\left(\frac{\sqrt{\beta^2+a'^2}}{b+\sqrt{c^2+a'^2}}\right).$$

If, in this value of F, we invert the ratios, in order to make the logarithms affirmative, and write like quantities, one under the other, we have

$$F = \eta a - \eta' a' + b\log \frac{\beta+\sqrt{c^2+a'^2}}{\beta+\sqrt{c^2+a^2}} \times \frac{\sqrt{b^2+a^2}}{\sqrt{b^2+a'^2}}$$
$$+\beta\log \frac{b+\sqrt{c^2+a'^2}}{b+\sqrt{c^2+a^2}} \times \frac{\sqrt{\beta^2+a^2}}{\sqrt{\beta^2+a'^2}}.$$

The first two terms of this expression deserve particular attention, as η is an arch, such that $\sin\eta = \sin BAE \sin BAC$; therefore, by what has been before demonstrated, η is the measure of the angular space subtended at A by the rectangle BD. The first term in the value of F, therefore, is the product of the distance AB, into the angular space subtended by the rectangle BD. In like manner, the second term, or $\eta' a'$, is the product of the distance AK, into the angular space subtended by the rectangle KF.

The relation of the quantities expressing the ratios, in the two logarithmic terms, will be best conceived by substituting for the algebraic quantities the lines that correspond to them in the diagram. Because $c^2 = b^2 + \beta^2 = EB^2 + BC^2 = EC^2$, therefore $c = EC$ or BD. So also, $c^2 + a^2 = BD^2 + BA^2 = AD^2$, because ABD is a right angle, &c. Thus,

$$F = \eta a - \eta' a' + BE \cdot \log \frac{(AF+FN)AE}{(AD+DE)AN} +$$

$$BC \cdot \log \frac{(AF+FM)AC}{(AD+DC)AM}.$$

This expression for the attraction of a parallelepiped, though considerably complex, is symmetrical in so remarkable a degree, that it will probably be found much more manageable, in investigation, than might at first be supposed. That it should be somewhat complex, was to be expected, as the want of continuity in the surface by which a solid is bounded, cannot but introduce a great variety of relations into the expression of its attractive force. The farther simplification, however, of this theorem, and the application of it to other problems, are subjects on which the limits of the present paper will not permit us to enter. The determinations of certain *maxima* depend on it, similar to those already investigated. It points at the method of finding the figure, which a fluid, whether elastic or unelastic, would assume, if it surrounded a cubical or prismatic body by which it was attracted. It gives some hopes of being able to determine generally the attraction of solids bounded by any planes whatever; so that it may, some time or other, be of use in the Theory of Crystallization, if, indeed, that theory shall ever be placed on its true basis, and founded, not on an hypothesis purely Geometrical, or in some measure arbitrary, but on the known Principles of Dynamics.

FINIS.

ON

THE PROGRESS OF HEAT

IN

SPHERICAL BODIES.

PROGRESS OF HEAT

IN

SPHERICAL BODIES.*

1. An argument against the hypothesis of central heat has been stated by an ingenious author as carrying with it the evidence of demonstration.

"The essential and characteristic property of the power producing heat, is its tendency to exist everywhere in a state of equilibrium, and it cannot hence be preserved without loss or without diffusion, in an accumulated state. In the theory of Hutton, the existence of an intense local heat, acting for a long period of time, is assumed. But it is impossible to procure caloric in an insulated state. Waving every objection to its production, and supposing it to be generated to any extent, it cannot be continued, but must be propagated to the contiguous matter. If a heat, therefore, exist-

---

* From the Transactions of the Royal Society of Edinburgh, Vol. VI. (1812.)—Ed.

ed in the central region of the earth, it must be diffused over the whole mass; nor can any arrangement effectually counteract this diffusion. It may take place slowly, but it must always continue progressive, and must be utterly subversive of that system of indefinitely renewed operations which is represented as the grand excellence of the Huttonian Theory."* "Again," he observes, in giving what he says appears to him a demonstration of the fallacy of the first principles of the Huttonian System, "it will not be disputed, that the tendency of caloric is to diffuse itself over matter, till a common temperature is established. Nor will it probably be denied, that a power constantly diffusing itself from the centre of any mass of matter, cannot remain for an indefinite time locally accumulated in that mass, but must at length become equal or nearly so over the whole."†

2. I must confess, notwithstanding the respect I entertain for the acuteness and accuracy of the author of this reasoning, that it does not appear to me to possess the force which he ascribes to it; nor to be consistent with many facts that fall every day under our observation. A fire soon heats a room to a certain degree, and though kept up ever so long, if its intensity, and all other circumstances

* Murray's System of Chemistry, Vol. III. Appendix, p. 49.
† Page 51.

remain the same, the heat continues very unequally distributed through the room; but the temperature of every part continues invariable. If a bar of iron has one end of it thrust into the fire, the other end will not in any length of time become red-hot; but the whole bar will quickly come into such a state, that every point will have a fixed temperature, lower as it is farther from the fire, but remaining invariable while the condition of the fire, and of the medium that surrounds the bar, continues the same. The reason indeed is plain: the equilibrium of heat is not so much a primary law in the distribution of that fluid, as the limitation of another law which is general and ultimate, consisting in the tendency of heat to pass with a greater or a less velocity, according to circumstances, from bodies where the temperature is higher, to those where it is lower, or from those which contain more heat, according to the indication of the thermometer, to those which contain less. It is of this general tendency, that the equilibrium or uniform distribution of heat is a consequence,—but a consequence only contingent, requiring the presence of another condition, which may be wanting, and actually is wanting, in many instances. This condition is no other, than that the quantity of heat in the system should be given, and should not admit of continual increase from one quarter, nor diminution from another. When such increase and diminution take place, what is usually called "the equilibrium of heat" no longer exists. Thus, if we expose a thermometer to the sun's rays, it immediately rises, and continues to stand above the temperature of the surrounding air. The way in which this happens is perfectly understood: the mercury in the thermometer receives more heat from the solar rays than the air does; it begins therefore to rise as soon as those rays fall on it; at the same time, it gives out a portion of its heat to the air, and always the more, the higher it rises. It continues to rise, therefore, till the heat which it gives out every instant to the air, be equal to that which it receives every instant from the solar rays. When this happens, its temperature becomes stationary; the momentary increment and decrement of the heat are the same, and the total, of course, continues constant. The thermometer, therefore, in such circumstances, never acquires the temperature of the surrounding air; and the only equilibrium of the heat, is that which subsists between the increments and the decrements just mentioned: these indeed are, strictly speaking, *in equilibrio*, as they accurately balance one another. This species of equilibrium, however, is quite different from what is implied in the uniform diffusion of heat.

3. In order to state the argument more generally, let A, B, C, D, &c. be a series of contigu-

ous bodies; or let them be parts of the same body; and let us suppose that A receives, from some cause, into the nature of which we are not here to inquire, a constant and uniform supply of heat. It is plain, that heat will flow continually from A to B, from B to C, &c.; and in order that this may take place, A must be hotter than B, B than C, and so on; so that no uniform distribution of heat can ever take place. The state, however, to which the system will tend, and at which, after a certain time, it must arrive, is one in which the momentary increase of the heat of each body is just equal to its momentary decrease; so that the temperature of each individual body becomes fixed, all these temperatures together forming a series decreasing from A downwards. To be convinced that this is the state which the system must assume, suppose any body D, by some means or other, to get more heat than that which is required to make the portion of heat which it receives every moment from C, just equal to that which it gives out every moment to E; as its excess of temperature above E is increased, it will give out more heat to E, and as the excess of the temperature of C above that of D is diminished, D will receive less heat from C; therefore, for both reasons, D must become colder, and there will be no stop to the reduction of its temperature, till the increments and decrements become equal as before.

4. If, therefore, heat be communicated to a solid mass, like the earth, from some source or reservoir in its interior, it must go off from the centre on all sides, toward the circumference. On arriving at the circumference, if it were hindered from proceeding farther, and if space or vacuity presented to heat an impenetrable barrier, then an accumulation of it at the surface, and at last a uniform distribution of it through the whole mass, would inevitably be the consequence. But if heat may be lost and dissipated in the boundless fields of vacuity, or of ether, which surround the earth, no such equilibrium can be established. The temperature of the earth will then continue to augment only, till the heat which issues from it every moment into the surrounding medium, become equal to the increase which it receives every moment from the supposed central reservoir. When this happens, the temperature at the superficies can undergo no farther change, and a similar effect must take place with respect to every one of the spherical and concentric strata into which we may conceive the solid mass of the globe to be divided. Each of these must in time come to a temperature, at which it will give out as much heat to the contiguous stratum on the outside, as it receives from the contiguous stratum on the inside; and, when

this happens, its temperature will remain invariable.

5. That we may trace this progress with more accuracy, let us suppose a spherical body to be heated from a source of heat at its centre; and let h, h', h'', be the temperatures at the surfaces of two contiguous and concentric strata, the distances from the centre being x, x', x''; and let it also be supposed, that the thickness of each of the strata, to wit, $x'-x$, and $x''-x'$, is very small.

Then supposing the body to be homogeneous, the quantity of heat that flows from the inner stratum into the outward, in a given time, will be proportional to the excess of its temperature above that of the outward stratum multiplied into its quantity of matter, that is, to $(h-h')(x'^3-x^3)$.

6. In the same manner, the heat which goes off from the second stratum in the same time, is proportional to $(h'-h'')(x''^3-x'^3)$; and these two quantities, when the temperature of the second stratum becomes constant, must be equal to one another, or $(h-h')(x'^3-x^3)=(h'-h'')(x''^3-x'^3)$.

But because $h-h'$, and $x'-x$ are indefinitely small, $h-h'=\dot{h}$, and $x'^3-x^3=3x^2\dot{x}$; therefore $\dot{h}\times 3x^2\dot{x}$ $=$ a given quantity; which quantity, since $\dot{x}$ is given, we may represent by $a^2\dot{x}^2$; so that $\dot{h}=\frac{a^2\dot{x}^2}{3x^2\dot{x}}$ $=\frac{a^2\dot{x}}{3x^2}$, or because $\dot{h}$ is negative in respect of $\dot{x}$, being a decrement, while the latter is an increment, $\dot{h}=-\frac{a^2\dot{x}}{3x^2}$, and therefore $h=C+\frac{a^2}{3x}$.

7. To determine the constant quantity C, let us suppose that the temperature at the surface of the internal *nucleus* of ignited matter is $=H$, and $r=$ radius of that nucleus. Then, in the particular case, when $x=r$ and $h=H$, the preceding equation gives $H=C+\frac{a^2}{3r}$; so that $C=H-\frac{a^2}{3r}$, and consequently $h=H-\frac{a^2}{3r}+\frac{a^2}{3x}$; or $h=H+\frac{a^2}{3}\left(\frac{1}{x}-\frac{1}{r}\right)$.

8. It is evident, from this formula, that for every value of x there is a determinate value of h, or that for every distance from the centre there is a fixed temperature, which, after a certain time, must be acquired, and will remain invariable as long as the intensity and magnitude of the central fire continue the same.

9. It remains for us to determine the value of a^2, which, though constant, is not yet given, or known from observation.

At the surface of the globe we may suppose the mean temperature to be known: let T be that temperature, and let $R=$ the radius of the globe. Then, when $x=R$, $h=T$, and by substituting in the general formula, we have $T=H+\frac{a^2}{3}\left(\frac{1}{R}-\frac{1}{r}\right)$, and $a^2=\frac{3(T-H)}{\frac{1}{R}-\frac{1}{r}}=\frac{3Rr(H-T)}{R-r}$.

Thus $h=\mathrm{H}+\frac{\mathrm{R}r(\mathrm{H}-\mathrm{T})}{\mathrm{R}-r}\left(\frac{1}{x}-\frac{1}{r}\right)=\mathrm{H}-\frac{\mathrm{R}r(\mathrm{H}-\mathrm{T})}{\mathrm{R}-r}$ $\left(\frac{1}{r}-\frac{1}{x}\right)$.

Hence also by reduction $h=\frac{\mathrm{RT}-r\mathrm{H}}{\mathrm{R}-r}+\frac{\mathrm{R}r(\mathrm{H}-\mathrm{T})}{x(\mathrm{R}-r)}$,

or $h=\frac{1}{\mathrm{R}-r}(\mathrm{RT}-r\mathrm{H})+\frac{\mathrm{R}r(\mathrm{H}-\mathrm{T})}{x}$.

From this equation, it is evident, that

$h-\frac{\mathrm{RT}-r\mathrm{H}}{\mathrm{R}-r}$, or the excess of the temperature at any distance x from the centre, above a certain given temperature, is inversely as x. But the construction of the hyperbola which is the locus of the preceding equation, will exhibit the relation between the temperature and the distance, in the way of all others least subject to misapprehension.

Let the circle (Fig. 37,) described with the radius AB, represent the globe of the earth; and the circle described with the radius AH an ignited mass at the centre. Let HK, perpendicular to AB, be the temperature at H, the surface of the ignited mass; and let FD be the temperature at any point whatever, in the interior of the earth, BM representing that at the surface. Then AB being $=\mathrm{R}$ in the preceding equation, $\mathrm{AH}=r$, $\mathrm{HK}=\mathrm{H}$, $\mathrm{BM}=\mathrm{T}$; $\mathrm{AF}=x$, and $\mathrm{FD}=h$, these two last being variable quantities; since

$\left(h-\frac{\mathrm{RT}-r\mathrm{H}}{\mathrm{R}-r}\right)x=\frac{\mathrm{R}r(\mathrm{H}-\mathrm{T})}{\mathrm{R}-r}$ we have, (taking AE

$=\frac{RT-rH}{R-r}$, and drawing EL parallel to AB, meeting HK in N, and FD in O,) $OD\times OE$ $=\frac{BA.AH(HK-BM)}{BH}$, which is a given quantity.

Therefore D is in a rectangular hyperbola, of which the centre is E, the asymptotes EG and EL, and the rectangle of the co-ordinates, equal to $BA.AH\times\frac{HK-BM}{BH}$, or, which amounts to the same, to KN.NE.

It is evident from this, that if the sphere were indefinitely extended, the temperature at the point B and all other things remaining the same, the temperature at its superficies would not be less than AE, or than the quantity $\frac{RT-rH}{R-r}$.

The quantity AE, or $\frac{RT-rH}{R-r}$, is supposed here to be subtracted; if RT be less than rH, it will change its sign, and must be taken on the other side of the centre A.

10. The results of these deductions may be easily represented numerically, and reduced into tables, for any particular values that may be assigned to the constant quantities. Thus, if the radius of the globe, or $R=100$, that of the ignited nucleus or $r=1$; the temperature of the nucleus, or $H=1000$, and T the temperature at the

surface $=60$, the formula becomes $h=50.505+\frac{949.494}{x}$.

Values of x	Values of h
10	$145^0.454$
20	98 .423
30	82 .599
40	74 .686
50	69 .938
60	66 .330
70	63 .926
80	62 .361
90	61 .055
100	60 .

11. Other things remaining as before, if we now make $r=10$, then $h=-44.444+\frac{10444}{x}$

x	h
20	$477^0.556$
30	303 .556
40	226 .556
50	164 .556
60	145 .556
70	104 .556
80	85 .056
90	64 .556
100	60 .000

12. If R=10, r=1, H=10000 and T=60

$$h = -1044.44 + \frac{11044.44}{x}$$

Values of x	Values of h
1	10000°.00
2	4477 .70
3	2637 .04
4	1716 .67
5	1164 .44
6	796 .30
7	533 .33
8	346 .16
9	182 .72
10	60 .00

13. 1. The general conclusions which result from all this are, that when we suppose an ignited nucleus of a given magnitude, and a given intensity of heat, there is in the sphere to which it communicates heat a fixed temperature for each particular stratum, or for each spherical shell, at a given distance from the centre; and that a great intensity of heat in the interior, is compatible with a very moderate temperature at the surface.

2. However great the sphere may be, the heat at its surface cannot be less than a given quantity; R, r, H and T, remaining the same. It must be observed, that though R is put for the radius of the globe; it signifies in fact nothing, but the distance at which the temperature is T, as r does the distance at which the temperature is H.

Therefore, were the sphere indefinitely extended, the temperature at its superficies would not he less than the quantity $\frac{RT-rH}{R-r}$, that is, not less than 50.5 in the first of the preceding examples, than —44.4 in the second, or —1044.4 in the third.

14. In all this the sphere is supposed homogeneous; but if it be otherwise, and vary in density, in the capacity of the parts for heat, or in their power to conduct heat, providing it do so as any function of the distance from the centre, the calculus may be instituted as above. For example, let the density be supposed to vary as $\frac{b}{b+x}$, then we have as before $(h-h')(x'^3-x^3)\frac{b}{b+x}$ for the momentary increment of heat in a stratum placed at the distance x from the centre, or $\dot{h}\times 3x^2\dot{x}\times\frac{b}{b+x}=$ to a given quantity, or to $a^2\dot{x}^2$, and therefore $\dot{h}=-\frac{a^2(b+x)\dot{x}}{3bx^2}=-\frac{a^2\dot{x}}{3x^2}-\frac{a^2\dot{x}}{bx}$.

Hence $h=C+\frac{a^2}{3x}-\frac{a^2}{b}\log x$. Suppose that when $x=r$ the radius of the heated nucleus, $h=H$; then H=

$$C+\frac{a^2}{3r}-\frac{a^2}{b}\log r, \text{ and } C=$$

$$H-\frac{a^2}{3r}+\frac{a^2}{b}\log r; \text{ therefore } h=$$

$$H-\frac{a^2}{3r}+\frac{a^2}{3x}+\frac{a^2}{b}\log\frac{r}{x}.$$

In this expression a^2 will be determined, if the

temperature at any other distance R from the centre is known. Let this be T; then by substitution we have

$$T = H - \frac{a^2}{3r} + \frac{a^2}{3R} + \frac{a^2}{b}\log\frac{r}{R},$$

$$\text{and } a^2 = \frac{T - H}{\frac{1}{3R} - \frac{1}{3r} + \frac{1}{b}\log\frac{r}{R}}.$$

$$\text{Hence } h = H + \left(\frac{T - H}{\frac{1}{3R} - \frac{1}{3r} + \frac{1}{b}\log\frac{r}{R}}\right) \times$$

$$\left(-\frac{1}{3r} + \frac{1}{3x} + \frac{1}{b}\log\frac{r}{x}\right).$$

This is given merely as an example of the method of conducting the calculus when the variation of the density is taken into account, and not because there is reason to believe that the law which that variation actually follows, is the same that has now been hypothetically assumed.

16. The principle on which we have proceeded, applies not only to solids, such as we suppose the interior of the earth, but it applies also to fluids like the atmosphere, provided they are supposed to have reached a steady temperature. The propagation of heat through fluids is indeed carried on by a law very different from that which takes place with respect to solids; it is not by the motion of heat, but by the motion of the parts of the fluid itself. Yet, when we are seeking only the mean result, we may suppose the heat to be so diffused,

that it does not accumulate in any particular stratum, but is limited by the equality of the momentary increments and decrements of temperature which that stratum receives. This is conformable to experience; for we know that a constancy, not of temperature, but of difference between the temperature of each point in the atmosphere, and on the surface, actually takes place. Thus, near the surface, an elevation of 280 feet produces, in this country, a diminution of one degree. The strata of our atmosphere, however, differ in their capacity of heat, or in the quantity of heat contained in a given space, at a given temperature. Concerning the law which the change of capacity follows, we have no certain information to guide us; and we have no resource, therefore, but to assume a hypothetical law, agreeing with such facts as are known, and, after deducing the results of this law, to compare them with the observations made on the temperature of the air, at different heights above the surface of the earth.

17. Let us then suppose, that the strata of the atmosphere have a capacity for heat, which increases as the air becomes rarer, so as to be proportional to $mb^{-\frac{r}{x}}$, x denoting, as before, the distance from the centre of the earth, r the radius of the earth, m and b determinate, but unknown quantities, such that mb^{-1} or $\frac{m}{b}$, expresses the capacity of air

for heat, when of its ordinary density, at the surface of the earth. The formula thus assumed, agrees with the extreme cases; for, when $x=r$, the capacity of heat $=\frac{m}{b}$, a finite quantity; when x increases, $\frac{r}{x}$ diminishes, and so also does $b^{\frac{r}{x}}$, if b is greater than unity, and therefore $\frac{m}{b^{\frac{r}{x}}}$ increases continually. It does not, however, increase beyond a certain limit, for when x is infinite $\frac{m}{b^{\frac{r}{x}}}$ becomes $\frac{m}{1}$ or m.

18. Hence, by reasoning as in § 6, the momentary increment of the temperature, or sensible heat, of any stratum, is as $-\frac{a^2\dot{x}}{3x^2}$ directly, and its capacity for heat, or $mb^{-\frac{r}{x}}$ inversely, that is, $\dot{h}=-\frac{a^2\dot{x}}{3x^2}\times\frac{b^{\frac{r}{x}}}{m}=-\frac{a^2b^{\frac{r}{x}}\dot{x}}{3mx^2}$.

Let $\frac{r}{x}=y$, then $-\frac{r\dot{x}}{x^2}=\dot{y}$, so that $-\frac{a^2\dot{x}}{3mx^2}=\frac{a^2\dot{y}}{3mr}$, and therefore $\dot{h}=\frac{a^2b^y}{3mr}\dot{y}$. Hence $h=C+\frac{a^2}{3mr\log b}b^y=C+\frac{a^2}{3mr\log b}b^{\frac{r}{x}}$.

19. To determine C, if T be the temperature of the air at the surface, when $x=r$, $T=C+\frac{a^2b}{3mr\log b}$, and $C=T-\frac{a^2b}{3mr\log b}$.

Hence $h=T-\frac{a^2b}{3mr\log b}+\frac{a^2b^{\frac{r}{x}}}{3mr\log b}=T-\frac{a^2(b-b^{\frac{r}{x}})}{3mr\log b}$.

This formula, when $x=r$, gives $h=T$, and when x is infinite, it gives $h=T-\frac{a^2(b-1)}{3mr\log b}$. In all intermediate cases, as x is greater than r, $b^{\frac{r}{x}}$ is less than b, (b being a number greater than 1,) and therefore $b-b^{\frac{r}{x}}$ is positive, so that h is less than T, as it ought to be.

20. We may obtain an approximate value of this formula, without exponential quantities, that will apply to all the cases in which x and r differ but little in respect of r, that is, in all the cases to which our observations on the atmosphere can possibly extend.

If, in the term $b^{\frac{r}{x}}$ we write $r+z$ for x, z being the height of any stratum of air above the surface of the earth, we have $b^{\frac{r}{x}}=b^{\frac{r}{r+z}}$.

21. But, from the nature of exponentials, we know

$b^{\frac{r}{x}} = 1 + \frac{r}{x}\log b + \frac{r^2(\log b)^2}{2x^2} + \frac{r^3(\log b)^3}{2.3x^3}$, &c $= 1 + \frac{r}{r+z}\log b + \frac{r^2(\log b)^2}{2(r+z)^2} +$, &c.

Now $\frac{r}{r+z} = 1 - \frac{z}{r} + \frac{z^2}{r^2} -$, &c. And if we leave out the higher powers of z, we have nearly

$$\frac{r}{r+z} = 1 - \frac{z}{r}$$

$$\frac{r^2}{(r+z)^2} = 1 - \frac{2z}{r}$$

$$\frac{r^3}{(r+z)^3} = 1 - \frac{3z}{r}, \text{ \&c.}$$

Therefore, by substitution, we have $b^{\frac{r}{r+z}} =$

$1 + \left(1 - \frac{z}{r}\right)\log b + \left(1 - \frac{2z}{r}\right)\frac{(\log b)^2}{2} +$, &c. $=$

$$\left\{\begin{array}{l} 1 + \log b + \frac{(\log b)^2}{2} + \frac{(\log b)^3}{2.3} +, \text{ \&c.} \\ -\frac{z}{r}\log b - \frac{z}{r}(\log b)^2 - \frac{z}{r}\left(\frac{(\log b)^3}{2}\right), \text{ \&c.} \end{array}\right\}$$

Now, from the nature of exponentials,

$$b = 1 + \log b + \frac{\log b^2}{2} + \frac{\log b^3}{2.3} +, \text{ \&c.}$$

And $\frac{z}{r}\log b + \frac{z}{r}(\log b)^2 + \frac{z}{r}\frac{(\log b)^3}{2}$, &c.

$= \frac{z}{r}\log b\left(1 + \log b + \frac{(\log b)^2}{2} +, \text{ \&c.}\right) = \frac{z}{r} b \log b$; therefore

when z is very small, $b^{\frac{r}{r+z}} = b - \frac{zb}{r}\log b$, and therefore

(§ 19), $a^2\frac{(b-b^{\overline{r+z}})}{3mr\log b}=a^2\frac{\left(b-b+\frac{bz\log b}{r}\right)}{3mr\log b}=\frac{a^2bz}{3mr^2}$; hence when z is very small, $h=\mathrm{T}-\frac{a^2bz}{3mr^2}$.

22. Therefore when z, or the height above the surface is small, h diminishes in the same proportion that the height increases, which is conformable to experience.

In our climate, when $z=280$ feet, $\frac{a^2b\times 280}{3mr^2}=1^\circ$; so that the co-efficient $\frac{a^2b}{3mr^2}=\frac{1}{280}$, and therefore $h=\mathrm{T}-\frac{z}{280}$.

When the constant quantities are thus determined, the formula agrees nearly with observation. In the rule for barometrical measurements, it is implied, that the heat of the atmosphere decreases uniformly; but the rate for each particular case is determined by actual observation, or by thermometers observed at the top and bottom of the height to be measured.

FINIS.

LITHOLOGICAL SURVEY

OF

SCHEHALLIEN.

LITHOLOGICAL SURVEY

OF

SCHEHALLIEN.*

THE astronomical observations made on the mountain Schehallien, in 1774, were confessedly of great importance to science. They ascertained the power of mountains to produce a sensible disturbance in the direction of the plumb-line; of consequence, they proved the general diffusion of gravity through terrestrial substances, and afforded data for determining the medium density of the earth, compared with that of the bodies at its surface.

The skill with which this very delicate experiment was conducted by Dr Maskelyne, and the ingenuity with which the results were deduced by Dr Hutton, were worthy of the objects in view, and of the reputation which these distinguished men have acquired in their respective departments of the mathematical sciences.

* From the Philosophical Transactions of the Royal Society of London, Vol. CI. (1811.)—ED.

One thing only seemed wanting to give to the determination of the earth's density all the accuracy that could be obtained from a single experiment, namely, a more precise knowledge of the specific gravity of the rock which composes the mountain, as being the object with which the mean density of the earth was immediately compared. The specific gravity of that rock was assumed to be to that of water as 5 to 2; which, though it be nearly a medium, when stones of every kind, from the lightest to the heaviest, are included, is certainly too small for Schehallien, the rocks of which belong to a class of a specific gravity considerably above the mean. The uncertainty arising from this source might not be of great amount, yet it was desirable that the quantity, or, at least, the limits of it should be accurately ascertained. In this light I knew, from repeated conversations, that the matter was regarded by both the gentlemen above named.

I had therefore long wished to attempt such a survey of the mountain as might afford a satisfactory solution of this difficulty; and having mentioned the circumstances to the Right Honourable Lord Webb Seymour, he entered readily into a scheme, which, without the assistance of his skill and activity, I should have been quite unable to carry into execution.

Accordingly, in June 1801, we took up our residence in a small village, as near as we could to the bottom of the mountain, and began our mineralogical survey, the result of which we think it our duty to submit to the Society, under the auspices of which the original experiment was undertaken.

It was obvious, that our first object must be to obtain specimens of all the varieties of rock in the mountain, which had any considerable difference in their external characters. These specimens must be such as had not been exposed to the action of the weather, were perfectly sound, with a fresh fracture, and taken from the living rock. In order to procure these, we soon found that it was not necessary to dig into the mountain or to blast the stones with gunpowder, for the native rock breaks out on the bare and rugged surface in abundance of places, and is so deeply intersected by the streams that it was easy, by the assistance of the hammer only, to procure specimens having all the conditions requisite for our purpose.

Supposing, however, that all this was accomplished, it would be insufficient to determine the mean density of the mountain, unless the quantity of rock of each particular kind could also be estimated; at least nearly. It was necessary, therefore, to know what proportion of the mountain consisted of one species of rock, and what of another, without which the average could not be determined.

Had the mean density been the only thing wanted, it would have been sufficient to know the quantity of each variety of rock; but in the search we were engaged in, it was necessary to know not only the quantity, but the position of each of these varieties, relatively to the observatories on the south and north faces of the mountain. This will be evident, when it is considered that it was the effect of each portion of the rock on the plumb-line in these observatories, that was the thing to be found, and that this effect must vary not only with the density of the rock, but with its distance from the observatory, and its obliquity in respect of the meridian. The mean density would therefore be insufficient for estimating the attraction of the mountain, could it be found ever so exactly; and it is easy to show, that while the mean density of a heterogeneous mass, and also its magnitude and figure remain the same, its attractive force at a given point may be greatly changed by a different distribution of the materials it consists of, relatively to that point. In order then to form an estimate of this attraction, we must know, at least nearly, these three things, the varieties of rock composing the mountain; the quantity of each variety; and, lastly, the position of each relatively to the observatory. Fortunately the geometrical survey of the mountain, which had already determined not merely its superficial extent, but its solidity, taken in combination, with some peculiarities in its structure, has enabled us to approximate, I hope with some tolerable exactness, to the knowledge of all these three circumstances.

The plan, then, which we proposed to follow, and which was necessary to be pursued if our lithological survey was to correspond in any degree to the accuracy of the geometrical survey, made under the direction of the Astronomer Royal, was to try to recognise the chain of stations which had been employed in that survey, in order that, by reference to those stations, we might be able to determine the points on the surface of the mountain from which our different specimens were collected. After these stations were discovered, we meant to traverse the mountain in various directions, and at any point where a specimen was taken, to determine our position by the bearings of any two of the stations that might be in sight, or by taking angles to three of them, or such other methods as occasion and circumstances might suggest. This was to be done where considerable variations in the external characters of the rocks gave reason to look for considerable variations of specific gravity. It was an operation that could not be necessary for every individual specimen, but it was one which must be necessary for determining the district over which stone of a particular character prevailed. In this part of the work, we were to employ a theodolite, a sextant, or a compass, according as more or less accuracy seemed requisite.

As the marks of these stations were all effaced except some traces of the observatories, (or rather the huts in which Dr Maskelyne had lived,) and the two cairns on the top of the mountain, the discovery of the whole chain was a matter of some difficulty. By means, however, of the bearings, as given in Dr Hutton's paper, and the assistance of one of the guides who had been employed about the survey, we succeeded in finding out the stations; and as they were mostly on elevated points, we could distinguish them at a distance with sufficient exactness.

Schehallien belongs to one of the central ridges of the Grampians, which, stretching here from about SE. to NW. divides the vallies of the Tummel and the Tay. Though it be a part of this chain, it stands considerably separate from the rest on a base of a form somewhat oval, and having its figure distinctly defined by two streams that run, the one on the south, and the other on the north side of it. The lowest point in this base, which is on the NE. is 2467 feet below the summit of the mountain, and about 1094 above the level of the sea.

At the NW. extremity, Schehallien adheres to the main chain by means of a high ridge, depressed at its lowest point, little more than 1500 feet under the summit. On the opposite sides of this neck the streams rise, which were before said to determine the base of the mountain; these streams, however, do not unite at the eastern extremity of the base, for there also a sort of neck, though very low in comparison of the former, connects Schehallien with the hills to the eastward.

Beyond the streams just mentioned, a range of inferior hills, some of them very low, springing from the main ridge on the NW. encompasses the mountain, forming as it were a line of circumvallation round it, and on these were the stations which Mr Burrows, under Dr Maskelyne's direction, had chosen for the survey. Beyond these hills the ground falls down into a sort of plain of great extent on the north; on the south, less considerable and more uneven, yet such as to leave Schehallien very free and open in the direction of the meridian, and adapted by that means to show the full amount of its action on the plummet. From the base its sides rise with a rapid, though unequal acclivity, and terminate not in a point, but in a ridge or narrow plane of a waving form, about a mile in length, and sloping regularly to the east, where it is 480 feet lower than at the western extremity. Though the sides are very rugged, they are less broken by deep ravines or bold projections, than the other mountains of the same elevation in this quarter of the Grampians; for, beside the high neck which has been already mentioned as uniting Schehallien to the mountains on the west, it has only one other salient ridge, which runs out to the NE. and overlooks the plain with a very steep and precipitous aspect. In some directions, and when viewed from a considerable distance, the harsh features of the mountain are wonderfully softened; it acquires a very beautiful conoidal shape, and from thence derives the name by which it is known among the inhabitants of the low country.

The rock of Schehallien, like that of all the mountains in its vicinity, is of the class called primitive; and is disposed for the most part in great parallel plates, or strata, nearly vertical, stretching from SE. to NW. They are indeed so nearly vertical, that a deviation of 15° from the perpendicular is rarely to be met with, except toward the base of the mountain, when it is sometimes greater, and is subject to considerable inequalities. The strata on the north side of the mountain lean a little toward the north, and those on the south toward the south. All these variations, however, are inconsiderable, and in general the strata may be set down as nearly vertical.

But though in their disposition all the rocks of Schehallien agree pretty nearly, they differ considerably in their mineralogical characters. A large proportion of the mountain, and that which constitutes the most elevated part, is formed of a granular quartz, extremely hard, compact, and homogeneous. The whole mass from about the level of the two observatories up to the summit of the mountain, is of this stone. Lower down, again, on every side, the rock is a schistus containing much mica and hornblende; and the division into parallel and vertical plates is more obvious than in the granular quartz. This last, however, is sometimes found in the lower parts, forming thin, vertical plates, interstratified with the hornblende and mica slate, and all together preserving their parallelism with a neatness and accuracy which a work of art could hardly exceed. This is particularly to be observed in the bed of the *burn* of Glenmore, the stream that defines the base of the mountain on the south, and which toward the lower part of its course intersects the strata to a great depth.

Besides these two kinds of rock, we meet in several places toward the base of the mountain with a granular and micaceous limestone, highly crystallized, which, in one or two places, ascends to a considerable height. All these rocks are disposed in strata; but there are also veins or dikes of porphyry and grünstein, which traverse the mountain in different directions; one of the former kind, of great breadth, cuts it right across from NE. to SW., not far from the point of its greatest elevation. There is nowhere any appearance of metallic veins.

The quartzy rock of Schehallien is extremely hard and homogeneous, and by its slow and uniform decay, has no doubt given rise to that massive shape, and comparatively unbroken surface, which have been already remarked. Yet even here the work of time is abundantly evident; for the rock being much cut by fissures tranverse to its stratification, it separates and falls down in large prismatic fragments. Some of these are of a vast size, and being extremely durable, the accumulation of them where the ground is not too steep to permit them to lie, is very great, so that large tracts of the sides of the mountain are covered with cubical blocks of granular quartz, resting on one another, and *steadied* only by their own weight.

It is remarkable of the quartzy stone, that when exposed for some time to the weather, it acquires the lustre and appearance of white enamel, so that the old weather-beaten surface is more clear and shining than that which is immediately produced from a fresh fracture. The reason seems to be, that the stone does not consist of pure quartz, but along with the grains of quartz has a great number of grains of feldspar interspersed, which, when it is first broken, give it an opaque and earthy appearance. These are soon dissolved by the action of the weather; and there is then left over the surface a coat of pure quartz, which has the semi-transparency and vitreous gloss belonging to enamel.

The feldspar which enters into the composition

of the rock here described, is not always in grains, but in some specimens is found regularly crystallized. The crystals, however, are small, and very thinly disseminated; were they in more abundance, this stone might be accounted a granite, as Professor Jameson has remarked of a stone of the same kind which he found in the island of Jura.

From the vertical position of the strata we may infer with some probability, that the rock which breaks out anywhere at the surface, continues the same through the interior of the mountain, in the direction of a perpendicular plane, down to its base, or perhaps to an indefinite depth. The same stratum usually remains of the same nature to a great extent, whenever we have an opportunity of examining it, whether in a horizontal or a perpendicular direction; and it is not to be doubted that the same holds when no such opportunity occurs. When, therefore, we have on the surface a bed of mica slate, or of granular limestone, or of granular quartz, the probability is, that the whole stratum all through the mountain is composed of the same materials. I must, however, confess, that I do not think that this probability is as strong with respect to granular quartz, as it is with respect to the micaceous rocks. These last compose the great mass of the Grampians; and their characters, though not everywhere the same, change very slowly, and pass from one to another by imperceptible gradations. To the granular quartz this rule does not equally apply; it is nòt general among the mountains of this tract; it sometimes breaks off suddenly, and is replaced by rocks of a very different nature. We cannot, therefore, with the same confidence assume the existence of this rock in intermediate points, when we only see it in the extremes. This much, however, we know with certainty, that the whole of the upper part of the mountain, from about the level of the south observatory to the summit, consists of granular quartz, as no other stone is to be met with anywhere in that tract. This is the part above O, in the section of the mountain; and the only question is, whether we shall consider the part in the interior of the mountain, immediately under this mass, as consisting of the same rock. When we first examined Schehallien, Lord Webb Seymour and I were both of opinion that this was the most probable supposition. Since that time, however, having had an opportunity of examining some other of the Grampians where granular quartz is found at the summit, and where, nevertheless, it is certain that the same rock does not go down into the interior, there has appeared some reason to suspect that this may be true of Schehallien. As the result of the calculus with regard to the earth's density is materially affected by these suppositions, I have given the result as I had first deduced it on the hypothesis, that the interior of the mountain is of granular quartz ; and also on the hypothesis that the quartz is confined to the upper part ; and that the lower part is entirely composed of mica and hornblende slate.

In the computation which Dr Hutton made of the attraction of Schehallien, he supposed its mass divided into 960 vertical columns, and he computed the force with which each of these columns disturbed the direction of a plummet suspended in either observatory, supposing them all homogeneous, and two and a half times as dense as water. * Now, knowing from our survey, and the combination of geometrical with mineralogical observations, the specific gravity of each of these columns at the surface, and conceiving (what we have shown, with one exception, to be probable) that the column remains the same through its whole length, we can compare the real attraction with that assigned to it in Dr Hutton's calculation. The attraction of any column computed on his hypothesis being divided by 2.5, and multiplied by the true specific gravity, will give the real attraction, or effect in disturbing the plumb-line. It is on this principle that our correction is formed, though simplifications occurred that very much diminish-

* Phil. Trans. Vol. LXVIII. (1778,) p. 689, &c.

ed the labour of the computation, the nature of the rock leading us in the end to distinguish only two differences of specific gravity; and the ingenious deductions of Dr Hutton, together with the excellent order that prevails in his computation, having made it easy to follow a route which he had cleared of all its greatest difficulties.

However, as it was impossible to determine beforehand how much the specific gravity of these rocks might differ, it was necessary to conduct the survey so that every individual column, had it been necessary, might have had its specific gravity defined. For this purpose the mountain was traversed in various directions, and the points at which a transition was made from rocks of one character to those of another were carefully noted, and their position ascertained. In selecting the specimens which were to represent the rocks of the several districts into which the mountain thus became divided, attention was paid both to the prevailing stone, and to that which was least common, in order that we might, if possible, get possession both of the mean and the extremes. This was in general the principle that guided our choice of specimens, but the application of it in detail to particular instances does not admit of being explained. The reasons in every such case that determined us to take one stone and reject another, could only be perceived by an observer on the spot, whose eye was accustomed to judge of the varieties, the plenty or the rarity of the minerals that passed in review before him, by indications which it is impossible to describe in words. We are here therefore with reluctance compelled to request from the reader more credit than we are able to prove to him that we deserve. We know, that in doing this, we are craving an indulgence which no wise and candid observer ever wished to possess; we sincerely regret that the nature of the subject forces us to make this demand, and that the part of our work which it was most difficult to perform to our own satisfaction, is quite incapable of being explained to the satisfaction of others.

Catalogue of Specimens from Schehallien.

The rock of the mountain may be divided, as already remarked, into three classes; granular quartz; mica and hornblende slate; granular limestone. The specific gravities were ascertained by the late Dr Kennedy, and it is therefore unnecessary to add that their accuracy may be perfectly relied on. The pieces weighed were between 1000 and 4000 grains: most commonly between 2000 and 3000. Different pieces of the same specimen were often examined. The water used was distilled, and always of a temperature between 60 and 61 degrees.

Quartz.

1. Grey sandstone, containing mica in thin layers. Specific gravity =2.6435.

2. White quartz, very pure. Fracture vitreous. Occurs in beds chiefly on the NE. side of the mountain. Specific gravity =2.6437.

3. Quartzy sandstone, of a whitish grey colour, with thin layers of mica. Specific gravity =2.6296.

4. Quartzy sandstone. White colour, with layers of mica. Much indurated. Somewhat ferruginous. Specific gravity =2.65367.

5. Indurated sandstone, with spiculae of mica interspersed. Specific gravity =2.6460.

6. Sandstone, much indurated, vitreous shine, interspersed with mica. Specific gravity =2.6269.

7. Granular quartz from near the summit; contains grains of feldspar. Specific gravity =2.6274.

8. Granular quartz, nearly the same with the preceding. Specific gravity =2.6109.

9. Sandstone, fine grained, slightly marked with iron veins. Specific gravity =2.6296.

10. Sandstone, fine grained, more indurated than the preceding. Specific gravity =2.6576.

11. Sandstone containing calcareous matter. Specific gravity =2.6656.

12. Granular quartz, very compact and indurated, but of a stratified structure; a little mica

in thin plates. From a mean of several. Specific gravity =2.6452.

13. Granular quartz of a flesh colour; imper fect crystals of feldspar thinly disseminated. This specimen from near the top. From a mean of several pieces. Specific gravity =2.6387.

The mean of these thirteen specimens gives 2.6398 for the upper or quartzy part of the mountain.

Mica and Hornblende Slate.

1. Hornblende slate very compact. Specific gravity =3.0642.

2. Micaceous schistus, with hornblende, and a small mixture of quartz. Specific gravity =2.9385.

3. Black micaceous schistus, fine grained, containing hornblende. Specific gravity =3.0476.

4. Micaceous schistus, containing pyrites and quartz in fine grains. Specific gravity =2.7293.

5. Micaceous schistus tinged with an oxide of iron. Specific gravity =2.7935.

6. Micaceous schistus, with thin plates of mica and hornblende transverse to the stratification. Specific gravity = 907.

7. Micaceous schistus, with quartz in small grains. Specific gravity =2.7499.

8. Another specimen nearly the same. Specific gravity =2.7728.

9. Compact micaceous schistus, grains of feldspar and quartz intermixed. Specific gravity =2.71845.

10. Nearly the same with the preceding. Specific gravity =2.7206.

The medium specific gravity of these ten specimens is 2.83255.

Limestone.

1. Granular limestone of a grey colour, containing some mica. Specific gravity =2.7087.

2. Granular limestone, silver coloured, stratified structure. Specific gravity =2.8890.

3. The same, bluish, highly crystallized. Specific gravity =2.76057.

4. The same, finer grained, containing thin layers of mica. Specific gravity =2.7419.

5. The same, grey coloured, and the crystals larger. Specific gravity =2.7302.

The mean specific gravity of these five specimens is =2.76607.

Granular Quartz.		Micac. and Calc. Schistus.		
Numbers.	Specific Gravity.	Specific Gravity.	Numbers.	
1	2.6435	3.0642	1	Mic.
2	2.6437	2.9385	2	Mic.
3	2.6296	3.0476	3	Mic.
4	2.65367	2.7293	4	Mic.
5	2.6460	2.7935	5	Mic.
6	2.6269	2.7907	6	Mic.
7	2.6274	2.7499	7	Mic.
8	2.6109	2.7728	8	Mic.
9	2.6296	2.71845	9	Mic.
10	2.6576	2.7206	10	Mic.
11	2.6656	2.7087	1	Calc.
12	2.6452	2.8890	2	Calc.
13	2.6387	2.76057	3	Calc.
Mean	2.639876	2.7419	4	Calc.
		2.7302	5	Calc.
		2.81039	Mean	

From the inspection of the preceding table, it is evident that the specimens relatively to their specific gravity may be divided into two classes sufficiently distinct from one another. The specimens of granular quartz are in specific gravity comprehended between 2.61 and 2.66 nearly, and the mean is 2.639876. The micaceous rocks, including the calcareous, are contained between the limit 2.7 and 3.06, the mean of all the 15 specimens being 2.81039. Now it happens fortunately, that these two classes of rocks distinguished by their specific gravity are also distinguished by their position, so that the line which separates them can be accurately traced out on the face of the mountain.

As to the arrangement of the same two classes of rock in the interior of the mountain, there are only two different suppositions, as already observed, which possess any degree of probability, and the result of each is hereafter to be given. The curve line in the plan of the mountain divides the quartzy from the micaceous rocks.

I shall now proceed to state the principles on which the present investigation is founded, and the result to which it has led.

According to Dr Hutton's construction, if O (Fig. 38,) be the place of the plummet in the south observatory, ON the direction of the meridian, and if with a radius ON=13333 feet, or $\frac{40000}{3}$, a quadrant of a circle be described, viz. WRN; if ON be divided into 20 equal parts, and if from O as a centre, through each of these points of division, circles be described: lastly, if through O, radii as OH, OG, &c. be drawn such that the sine of the angle which each of them makes with the meridian shall differ from the sine of that which the contiguous radius makes with the meridian by $\frac{1}{12}$ of the radius; that is, if $\sin GON - \sin HON = \frac{1}{12}$, &c. then shall every one of the twenty concentric rings be divided into twelve spaces, upon each of which, if columns of homogeneous matter be supposed to stand, and to be of such altitudes as to subtend equal angles from O, the attraction of each column on the plummet at O, in the direction of the meridian ON, will be the same.

The attraction of any of these columns, as of that which stands on the base GHKL, is measured thus. Let b=GL, the breadth of the column in the direction of the radius, $=\frac{40000}{3.12}=666.666$ feet; $d=$ difference between the sines of the angles of azimuth, or sinGON—sinHON$=d$; E= angle of elevation of the column above O: then the attraction $=bd\sin E$.*

I have also used a theorem in these computations, which gives an accurate value of the attraction of a half cylinder of any altitude a, and any radius r, on a point in the centre of its base, and in the direction of a line bisecting the base. Let A be equal to that attraction; then $A=2a\log\frac{r+\sqrt{a^2+r^2}}{a}$, or $A=2a\log\frac{r}{a}\left(1+\sqrt{1+\frac{a^2}{r^2}}\right)$.

Fig. 39 represents a vertical section of Schehallien in the direction of the meridian of the south observatory O.† The line QR represents the level

---

* Phil. Trans. Vol. LXVIII. p. 751.

† The observatories O and P are not in the same meridian. They are, however, nearly so; and the section through P, in the direction of the meridian, would not differ sensibly from that which is here given.

of the lowest part of the base of the mountain. P is the north observatory; the part of the section coloured with a reddish brown represents the granular quartz, supposed here to constitute the interior as well as the summit of the mountain. The dark colour represents the schistus; the two belts of grey are the limestone strata on the north and south sides. OR, or the elevation of the south observatory above the lowest part of the base of the mountain is 1440 feet.

Fig. 40. is a section of the mountain in the direction perpendicular to the meridian of the south observatory. This section, though not referred to in any of the computations, is useful for enabling one to form an idea of the structure and figure of the mountain.

Draw OL (fig. 39.) parallel to the horizon. With OR as an axis, and with a radius of 13333 feet, suppose a cylinder to be described, and let it be cut into two semi-cylinders by a plane passing through OR perpendicular to RQ the meridian. Then the whole of the mountain on the north side of this plane disturbs the direction of the plummet by drawing it toward the north. But the part of the mountain to the north of this plane, and between the levels O and R is equal to one of the semi-cylinders above mentioned, *minus* the empty space between the surface of the ground and the horizontal plane passing through O. If, therefore,

S denote the attraction of the semi-cylinder, and V that which the void space would have were each pillar in it to consist of matter of the same density with the part of the same pillar which is under the surface, $S-V$ will represent the attraction or disturbing force of all that part of the mountain which is north of OR, and under the level of O.

Again, putting S' and V' to express the same things for the part of the mountain to the south of O, the whole attraction of that part equal $S'-V'$, and this acting in an opposite direction to the other, or tending to restore the plummet to its mean position, is to be subtracted from the former quantity, so that the whole disturbing force by which the part of the mountain below the level of O acts upon the plummet at O, is $S-V-S'+V'$. To this the attraction of the upper part of the mountain, or that which is above the level of O, being, as it happens, wholly to the north is to be added, and if it be called T, the whole disturbance on the plummet at O is $S-S'-V+V'+T$.

In Dr. Hutton's computation, S and S', or the attraction of the half cylinder on opposite sides of O are equal to one another, the cylinder being supposed to consist of matter of the same density throughout; they must therefore destroy one another, and consequently, according to that hypothesis, they did not require to be calculated. The case here is not the same; for the matter in the two semi-cylinders not being of uniform density, nor having its inequalities similarly distributed, the attraction of each must be calculated, in order that their difference, or S—S′, may be found.

If Σ, Σ', U, U′, and T′ denote the same quantities for the observatory P on the north side of the mountain, then the disturbing force on the plummet at P, $=\Sigma-\Sigma'-U+U'+T'$; and so the whole force which alters the direction of gravity is $S-S'-V+V'+T+\Sigma-\Sigma'-U+U'+T'$.

The computation of these quantities for the columns in the quadrant north-west of O, will serve to explain the method followed in all the rest.

The whole cylinder of which OR is the axis being divided into 960 columns, the quarter of it must consist of 240, all of which, as far as their bases are concerned, are of equal force in attracting the plummet at O, so that the difference of their effects depends entirely on their altitude. Let O, (Fig. 38.) represent the south observatory, ON the meridian, and the quadrant ONW a horizontal section through O of one-fourth of the cylinder, on which the bases of the columns are marked as in the figure. Let *abc* be the bounding line of the quartz projected on the plane of this section, the columns whose bases are within that line being supposed wholly of quartz, and those without it of micaceous schistus. If we suppose the columns that have their tops in this section to be extended downwards to the depth of 1440 feet, we shall have the quarter-cylinder divided into 240 columns, that would be of equal disturbing forces, were they of equal density, and equal apparent depression below the point O. The inspection of the figure serves to distinguish the columns of quartz from those of micaceous schistus. In those columns which consist of both rocks, the proportion of the quartzy to the micaceous part could be judged of with sufficient accuracy by the eye. To assist the eye, however, the figure being first constructed to a large scale, I used to stretch a fine thread either in the direction of a radius passing through O, or in a line at right angles to that direction, (according as the case seemed to require,) so as to divide the quadrilateral into two quadrilaterals, equal, as nearly as the eye could judge, to the irregular divisions made by the boundary of the quartz and schistus. The proportion of the parts was then easily ascertained. Now, by the first of the theorems laid down above, the attraction of any column on the plummet at O, estimated in the direction of the meridian ON, if b be the breadth of it in the direction of the radius, d the difference of the sines of the azimuths of the two edges, E the angle which the length of the column subtends at O, if its density were $=1$, would be $bd \times \sin E$. But if the density of the rock be expressed by any other number, the attraction just found must be multiplied by that number, in order to give A the real attraction of the column. Thus, if Q denote the density of the granular quartz, and M that of the micaceous schistus, we have in the former case $A = bdQ\sin E$, and, in the latter, $A = bdM\sin E$. In these formulas, $b = 666.66$ feet, and $d = \frac{1}{12}$, by the construction already explained; therefore $A = (55.55)Q\sin E$, or $= (55.55)M\sin E$.

The calculation of sinE is very easy, for the length of each column or its depth below O being 1440 feet, and the middle of the first ring being 333.33 feet distant from O; of the second 1000, reckoning from O, if n be the number of any ring, the distance of its centre from O is $\frac{2n-1}{3} \times 1000$, so that $\tan E = \frac{1440}{\frac{2n-1}{3} \times 1000} = \frac{3 \times 1.44}{2n-1}$:

The sine corresponding to this tangent taken from the tables, and multiplied into 55.55, and the product into Q or M, will give the attraction of the column. Therefore to have the attraction of the ring of columns of the order n, the quantity now obtained must be multiplied by 12, that being the number of columns in one ring, having all by hypothesis the same altitude, so that the whole attraction of the ring $= (666.66)Q\sin E$, &c.

The attraction of each of the twenty rings being thus computed, their sum gives the attraction of the quarter-cylinder. *

From the projection of the columns in Fig. 38. it appears that the first six rings in the NW. quadrant are entirely of quartz, that the five following are mixed, being partly quartzy partly micaceous, and that the nine remaining rings are wholly micaceous. The little table that follows contains the proportions of quartzy and micaceous rock in the five rings just mentioned.

Rings. \ Sectors	1	2	3	4	5	6	7	8	9	10	11	12
7	$\frac{7}{10}q$ $\frac{3}{10}m$	$\frac{8}{10}q$ $\frac{2}{10}m$	$\frac{9}{10}q$ $\frac{1}{10}m$	q	q	q	q	q	q	q	q	q
8	m	m	m	$\frac{1}{8}q$ $\frac{7}{8}m$	$\frac{2}{5}q$ $\frac{3}{5}m$	$\frac{8}{9}q$ $\frac{1}{9}m$	q	q	q	q	q	q
9	m	m	m	m	m	m	$\frac{2}{5}q$ $\frac{3}{5}m$	q	q	q	q	$\frac{7}{11}q$ $\frac{4}{11}m$
10	m	m	m	m	m	m	m	$\frac{1}{8}q$ $\frac{7}{8}m$	$\frac{7}{8}q$ $\frac{1}{8}m$	q	q	$\frac{1}{4}q$ $\frac{3}{4}m$
11	m	m	m	m	m	m	m	m	m	$\frac{1}{4}q$ $\frac{3}{4}m$	$\frac{1}{4}q$ $\frac{3}{4}m$	m

---

* It was most convenient to compute the attraction of the quarter-cylinder in this way, though merely an approximation, because the columns of which it consisted are not all of the same specific gravity. In the case of their being homogeneous, the attraction of the quarter-cylinder might be computed *exactly* by the second theorem given above. In-

This table is constructed only for that part of the north-west quadrant in which columns occur of two different rocks ; and a rectangular cell is assigned to each column in the five rings to which the table refers. The letters *q* and *m* denote quartz and mica ; and where one letter only occurs, the column is entirely of the rock which it denotes. In the cells where both letters occur, the column consists of both rocks in the proportion expressed by the fraction prefixed to each letter. Thus, in the seventh ring, the first quadrilateral is $\frac{7}{10}$ quartz and $\frac{3}{10}$ mica ; the second, $\frac{8}{10}$ quartz and $\frac{2}{10}$ mica ; the third, $\frac{9}{10}$ quartz and $\frac{1}{10}$ mica ; the remaining nine being entirely quartz.

Now to apply the tables thus constructed to the computation of the attraction of any of the quarter-cylinders, it must be observed, that sinE is to be found for any column in the way already explain-

---

deed, I investigated that theorem for the purpose of examining into the degree of accuracy that this approximation actually possessed ; and I had the satisfaction to find, that when the two methods were applied to the same half or quarter-cylinder, (supposed homogeneous,) the difference of the results did not exceed a two-thousandth part of the whole. This demonstrates in a very satisfactory manner the accuracy of the method pursued by Dr Hutton.

ed, and is then to be multiplied by bd, b being $=\frac{2000}{3}$ and $d=\frac{1}{12}$, so that $bd=\frac{2000}{3\times 12}=\frac{500}{9}$, and, therefore, the coefficient of Q or M is $\frac{500}{9}\sin E$.

When the whole ring is of the same rock, the coefficient of sinE computed for a single column is to be multiplied by 12, so that the whole attraction of the ring $=\frac{500}{9}\times 12=\frac{2000}{3}=666.66$, as before determined.

In the mixed columns the sine of E is to be multiplied both into bd, and into the fraction prefixed to q for the quartz, and to m for the mica; or if we would include the whole ring, as E is the same for all the columns contained in it, we must multiply bd by the numbers denoting the proportion of quartz or of mica in the whole of that ring. Thus in ring 7, the first in the preceding table, the whole quartz $=11.4$, the sine of E being $=.3157$, and the attraction due to the quartz $=\frac{11400.Q}{3\times 6}\times .3157=(129.9423)Q$.

In this manner the attraction of the whole cylinder on the plummet at O is readily computed; but it must be diminished on account of the part by which this cylinder rises above the surface of the ground. The quantity that is to be subtracted is computed from the sines of the depressions of the tops of the different columns below the observatory

O; and Dr Hutton's paper either actually exhibits * those sines, or furnishes us with the means of readily computing them.† When a ring is wholly of the same species of rock, the sum of the sines of the depression of all the columns in that ring is given in the tables, and needs only to be multiplied by $\frac{500}{9}$ to give the coefficient of Q or M as far as that ring is concerned.

Again, when in the same ring some columns are of quartz and others of mica, the sines of depression must be computed trigonometrically for each column by help of the data contained in the tables above referred to. The sum of those sines for the quartz columns being multiplied by bd, gives the coefficient of Q.

Where the same column is of two different kinds of rock, the sine of the depression, or of E, must be multiplied into bd, and divided in the proportion of the numbers prefixed to q and m in the cell belonging to the column.

All this may be illustrated by the calculation of the attraction of the columns belonging to the above table. In the seventh ring the first columns are mixed, the next three are entirely of quartz, and the remaining six are wanting, that is to say,

---

* Phil. Trans. Vol. LXVIII. p. 769 to 776.

† Ibid. p. 759 to 765.

their tops are not depressed below the level of O, as may be seen in Table V. of Dr Hutton's paper. From that table it also appears, that the depth of the summit of the first column of the seventh ring below the level of O is 250 feet; of the second 240, of the third 200, of the fourth 150, of the fifth 60, and of the sixth 30. From these measures the angles of depression may be computed. Thus, if 250 be divided by the radius of this ring, viz. $\frac{13000}{3}$, we have .0577 for the tangent of the depression, or of E, and the sine which corresponds is .0568. As $\frac{7}{10}$ of this column consists of quartz, we must take $\frac{7}{10}$ of this sine for the proportional part of the coefficient of Q. In like manner, the sine of the depression of the top of the second column is .0545, of which taking $\frac{8}{10}$ we get .0436 for the part of the coefficient of Q belonging to this column. So also for the third ring, the proportional part of the sine is .04149. The fourth, fifth, and sixth columns being entirely of quartz, no proportional parts are to be taken; their sines, computed as before, are .0346, .0138, .0069; and the sum of all these six numbers is .18015.

Calculating in the same way for all the columns that are entirely or partly of quartz in the north-west quadrant, we have the amount of the whole =.2534. Now the total sum of the sines of the depressions in this quarter, is 13.534. (See Dr Hutton's Computations, p. 83.) From this number, if .2534 be taken away, there will remain 13.2806 as the coefficient of M, arising from the depressions of the micaceous columns.

Now the sum of the sines belonging to the quartz in the quarter-cylinder itself has been found =53.3532, from which taking away .2534, there remains 53.0998 for the entire sum of the sines belonging to the quartz under the level of O in the north-west quarter of the moumtain.

In like manner the sum of the sines computed for the micaceous columns and parts of columns in the north-west quarter-cylinder, is 23.0124, from which taking away 13.2806, the deficient or negative part, there remains 9.7318. The numbers thus found being multiplied by $\frac{500}{9}$, give the coefficients of Q and M for the north-west quarter of the mountain below the level of O, and make its attraction =(2949.99)Q+(540.655)M.

A similar computation being made for the quarter-cylinder on the north-east of O, we have its attraction =(2974.299)Q+(577.98)M, to which adding their former attraction, (2949.99) Q+ (540.655)M, we have S—V, or the attraction of the mountain on the north side of O, and under the level of O =(5924.289)Q+(1118.635)M. In like manner the attraction of the south-west quadrant deduced partly from the quarter-cylinder, and partly from Dr Hutton's calculations =(1049.18) Q+(1819.66)M, and of the south-east =(1567.394) Q+(1052.129)M; the sum of which, or S′—V′, gives (2616.574)Q+(2871.789)M; to be subtracted from the former, in order that we may have the total disturbing force of the part of the mountain below O, which is therefore =(3307.715)Q—(1753.154)M.

Lastly T, or the attraction of the part of the mountain above O, (which is on the north,) when computed from the sums of the sines of the elevation of the columns above O, as given by Dr Hutton, is found =(2474.389)Q+(150.855)M, which, added to the preceding, gives the whole attraction on the plummet at O, =(5782.104)Q—(1903.209)M.

The same quantities calculated for P, the north observatory, are (8061.022)Q—(3127.05)M. To which adding the attraction just found for O, we have (13843.126)Q—(5030.214)M= the total force of attraction increasing the convergency of the plumb line on opposite sides of the mountain.

Now if D be the mean density of the globe, it follows from Dr Hutton's calculations that 87522720D is the measure of the attraction of the whole earth. But the Astronomer Royal having found by his observations, that the sum of the deviations of the plumb-line on opposite sides of the mountain is 11.6 seconds, the attraction of the earth is therefore to the sum of the opposite attractions of Schehallien, as radius to the tangent of 11″.6, that is, as 1 to .000056239, or as 17781 to 1; or, making an allowance for the centrifugal force arising from the earth's rotation, as 17804 to 1. Therefore, $17804:1::87522720D:(13843.126)Q-(5030.214)M$, so that $\frac{87522720}{17804}D=(13843.126)Q-(5030.214)M$, and hence

$$D=\frac{(13843.126)Q-(5030.214)M}{4915902}$$, or $D=(2.816)Q-(1.023)M$. If we suppose $Q=2.639876$ and $M=2.81039$, as in the table above, $D=4.55886$.

Dr Hutton makes $D=\frac{17804}{9933}$ multiplied into 2.5, the supposed density of the rock,* which gives $D=4.481$, considerably less than the preceding. If in the formula $D=(2.816)Q-(1.023)M$, we make $Q=M=2.5$, the result should agree with Dr Hutton's, and does so very nearly, making $D=4.482$.

In all this, we have proceeded on the supposition that the granular quartz not only constitutes the summit of the mountain, or the part above the level of the observatories, but that it also descends

* Phil. Trans. Vol. LXVIII. p. 781.

into the interior of the mountain down to its base, where it is bounded by the curve line *abc*, (Fig. 38.) On the other supposition mentioned above, that the granular quartz does not constitute the interior nucleus of the mountain, but is confined to the upper part of it, the rest consisting of micaceous schistus, our formula, after undergoing certain changes, may also be accommodated to this hypothesis. In the value of the attraction of the part of the mountain below O, viz. (3307.715)Q−(1753.154)M, we must suppose Q=M, when the above quantity becomes (1554.561)M. To this we are to add T, or the attraction of the part of the mountain above O, which remains the same as before, viz. (2474.389)Q−(150.055)M, to which if we add (1554.561)M, the sum (2474.389)Q+(1404.506)M is the whole attraction on the plummet at O, according to this new hypothesis.

If the same changes are made with respect to the observatory P, we shall have the total attraction. Now the attraction of the part of the mountain below P=(5593.347)Q−(3172.15)M, which if Q=M becomes (2421.197)M. Also the attraction of the part above P is (2467.675)Q+(45.15)M. If to this be added (2421.197)M, the amount, or (2467.675)Q+(2466.347)M is the total attraction on the plummet at P.

To the total attraction at O=(2474.389)Q+(1404.506)M, add the total attraction at P=

(2467.675)Q+(2466.347)M; then the total attraction by which the direction of gravity is altered by the mountain, is (4942.064)Q+(3870.853)M. Hence as before $D=\frac{(4942.064)Q+(3870.853)M}{4915902}$, or $D=(1.0053)Q+(0.78743)M$.

Here if we make as before Q=2.639876, and M=2.81039, we shall have D=4.866997. This therefore is the mean density of the earth, on the supposition that the interior of Schehallien, on a lower level than the observatories, consists of micaceous schistus. The measure thus obtained, for the mean density or mean specific gravity of the earth, is above that of any of the precious stones, and is nearly a mean between the results of Dr Hutton and Mr Cavendish. According to the former, D=4.481; according to the latter, D=5.48, the mean of which is 4.98. The difference between this and the last of our results is nearly =.1, or less than a forty-fifth part.

If we are to consider the experiments on Schehallien singly, it seems highly probable that the mean density of the earth is contained between the limits deduced from the two different suppositions concerning the structure of the mountain, so that it cannot be less than 4.5588, nor greater than 4.867. The mean of these is nearly 4.713.

It is however desirable, that an element so important in physical astronomy, as the mean density of the earth, should be the result of many experiments. The principle on which those at Schehallien were made seems the most likely to lead to accurate conclusions. In the selection of the places fit for such observations, the homogeneity of the rock is a condition that merits particular attention, and is hardly to be looked for any where but among granite mountains, as they alone afford a perfect security that their interior and exterior are composed of the same materials. Granite is the lowest of the rocks, and whenever it appears at the surface we may be assured, that on penetrating deeper, we shall meet with no other.

It is therefore to the primitive mountains, and among them to the granitic, that such experiments as those made at Schehallien ought to be confined. The want of homogeneity will then be on the outside of the mountain only, and can easily be estimated. The granite may be covered at the bottom of the mountain and even to a considerable height on its sides with beds of gneiss, mica slate, hornblende slate, &c. the quantity and position of which can easily be ascertained by observation.

FINIS.

ON THE

NAVAL TACTICS

OF THE LATE

JOHN CLERK, Esq. of Eldin.

MEMOIR

RELATING TO

NAVAL TACTICS.*

* * * * * * * * * *

THE author of the *Naval Tactics* is one of those men who, by the force of their own genius, have carried great improvements into professions which were not properly their own. The history both of the sciences and of the arts furnishes several remarkable examples of a similar nature. Fermat the rival, sometimes the superior of Descartes, one of the most inventive mathematicians of a most inventive age, was by profession a lawyer, and had only devoted to science the time that could be spared from the duties of a counsellor or a judge: about fifty years earlier, also, his countryman Vieta had made a like digression from the same employment, and hardly with inferior success.

---

* Published in the Transactions of the Royal Society of Edinburgh, Vol. IX. (1821.)—ED.

Perrault, who, in the façade of the Louvre, has left behind him so splendid a monument of architectural skill and taste, was a physician, and not only practised, but wrote books on medicine. Dr Herschell too, who has made more astronomical discoveries than any individual of the present age, betook himself to the study of the heavens as a relaxation from professional pursuits. Mr Clerk is to be numbered with these illustrious men, having made great improvements in an art to which he was not educated, and in which early instruction, and long practice, would seem more indispensable than in any other.

Two reasons may perhaps be assigned for the success which often attends men who thus take a science by assault, without making their approaches regularly, and according to the rules of art. They are inspired by genius, and impelled by the highest of all motives, the pleasure they derive from their exertions. They are also free from the prejudices, and the blind respect for authority, which constitute so strong a barrier against improvement both in the sciences and the arts.

A young man, who had been bred in the service of the Navy, who had seen the Commanders he was taught to respect most highly, bring their fleets into action constantly in a certain way, and who had naturally made that manœuvre the great object of his study, would not be apt to deviate from a practice, in the accurate and successful application of which the greatest merit of a Naval Officer was supposed to consist. Indeed no man learns his art, as it actually exists, more completely than a seaman ; but no man learns it in a way more likely to preclude improvement. A landsman, therefore, sitting in his study, and thinking only of the abstract principles, mechanical or tactical, of the naval art, provided he be well instructed in them, and have a mind sufficiently powerful to combine those principles, and appreciate their different results, may be expected to give valuable lessons to the most able and experienced seamen.

Mr Clerk was precisely the man by whom a successful inroad into a foreign territory was likely to be made. He possessed a strong and inventive mind, to which the love of knowledge, and the pleasure derived from the acquisition of it, were always sufficient motives for application. He had naturally no great respect for authority, or for opinions, either speculative or practical, which rested only on fashion or custom. He had never circumscribed his studies, by the circle of things immediately useful to himself ; and I may say of him, that he was more guided in his pursuits, by the inclinations and capacities of his own mind, and less by circumstance and situation, than any man I have ever known. Thus it was, that he studied the surface of the land as if he had been a general, and the surface of the sea as an admiral, though he had no direct connection with the profession either of the one or of the other.

From his early youth, a fortunate instinct seems to have directed his mind to naval affairs. It is always interesting to remark the small and almost invisible causes from which genius receives its first impulses, and often its most durable impressions. "I had (says he) acquired a strong passion for nautical affairs when a mere child. At ten years old, before I had seen a ship, or even the sea at a less distance than four or five miles, I formed an acquaintance at school with some boys who had come from a distant sea-port, who instructed me in the different parts of a ship from a model which they had procured. I had afterwards frequent opportunities of seeing and examining ships, at the neighbouring port of Leith, which increased my passion for the subject; and I was soon in possession of a number of models, many of them of my own construction, which I used to sail on a piece of water in my father's pleasure-grounds, where there was also a boat with sails, which furnished me with much employment. I had studied *Robinson Crusoe*, and I read all the sea-voyages I could procure."

The desire of going to sea, which could not but arise out of these exercises, was forced to yield to family considerations; but, fortunately for his country, the propensity to naval affairs, and the plea sure derived from the study of them, were not to be overcome.

He had indeed prosecuted the study so far, and had become so well acquainted with naval affairs, that, as he tells us himself, he had begun to study the difficult problem of the way of a ship to windward. This was about the year 1770, when an ingenious and intelligent gentleman, the late Commissioner Edgar, came to reside in the neighbourhood of Mr Clerk's seat in the country. Mr Edgar had served in the army, and, with the company under his command, had been put on board Admiral Byng's ship at Gibraltar, in order to supply the want of marines; so that he was present in the action off the Island of Minorca, on the 20th of May 1756. As the friend of Admiral Boscawen, he afterwards accompanied that gallant officer in the more fortunate engagement of Lagos Bay.

After the American war was begun, an attention to the narratives of his friend, and still more to the actions which were then happening at sea, served to convince Mr Clerk that there was something very erroneous in the methods hitherto pursued by the British admirals for bringing their fleets into action; insomuch, that, though nothing could exceed the skill with which each individual ship was worked, yet when one whole fleet was opposed to another, the plan followed was uncertain and precarious, and it seemed that the expedient for forcing an enemy to fight, remained yet to be discovered. It appeared, indeed, that very little attention had yet been paid to the subject of Naval Tactics.

The oldest work we know of that treats of Naval Tactics, is that of the learned Jesuit Pere Hoste, Professor of Mathematics in the Royal College at Toulon, and entitled *L'Art des Armees Navales.* It is an elementary and distinct exposition of the ordinary manœuvres at sea, and has no pretensions to any thing more. It was, however, highly regarded at the time: the author, when he presented it to Louis XIV. in 1697, was well received, and had a pension given him.

There was no book on the subject in the English language; and the conduct of our sea-fights, though it had so often proved successful, did not display much extent or variety of resources. It had usually happened that the British fleet was eager to engage, and that the enemy was unwilling to risk a general action; our object, therefore, had almost always been to gain the *weather-gage,* as it is called, of the enemy, or to place ourselves to windward of his fleet. When that fleet was drawn out in line, in the manner necessary for allowing every ship its share in the action, the British fleet bore down from the windward on the enemy, *lying to,* as it is termed, or *almost fixed in its position;* the whole line, and also the broadside of each individual ship, being nearly at right angles to the direction of the wind.

Under these circumstances, the British fleet had usually pursued one of two methods of making the attack. The one consisted in forming a line parallel and directly opposite to that of the enemy; after which each ship bore down on that which was immediately opposed to *it*. According to the other method, the British fleet, on the opposite tack to that of the enemy, ran along parallel to it, and within fighting distance, till the whole of the one line was abreast of the other, and each ship ready to engage her antagonist.

If the first of these methods was pursued, each ship, on coming down, had a very sharp fire to sustain from the broadside of that opposed to her, which she could only return feebly from the *guns on her bow*. The rigging, of consequence, which presented the best mark when seen endwise, was likely to be so much cut, that the ship must be nearly disabled before she arrived at the fighting distance.

If the second method was pursued, the headmost ship had to endure the fire of the whole line before it arrived in its place; the second, the fire of all but one; the third, of all but two, and so on: so that it was not likely that any but the sternmost ships could reach their station without having received considerable damage. This gave to the enemy who quietly remained on the defensive, a great advan. tage over the attacking squadron, and enabled him, almost to a certainty, to come off, if not with victory, at least with very little loss.

The disadvantages, however, arising from these two modes of attack, either had not been duly considered, or had been set down among the unavoidable evils necessarily involved in a determination to force the enemy to fight. Perhaps, too, the desire of complying literally with the instructions always given to our admirals, of doing their utmost to take, burn, and destroy, contributed to make it be thought, that a direct and immediate attack, such as has just now been described, was the only means that could properly be resorted to.

Mr Clerk had the merit of pointing out the evils now enumerated, in a manner most clear and demonstrative, and of describing a method by which the attack might be made, without incurring any of the disadvantages that have been mentioned, and almost with a certainty of success. As the evil arose from an endeavour to diffuse the force of the attack, if one may say so, over the whole surface of the line attacked; so the remedy consisted in concentrating the force of the attack, and in bringing it to bear with proportionally greater energy on a single point, or a small portion of the enemy's line. For this purpose the admiral of the attacking and windward squadron, is supposed to come down, not in line, but with his fleet in divisions, so as to be able to support the particular division destined to break through the line of the enemy. The consequence must be, that, if this attack is directed against the rear of the enemy, the ships a-head must either abandon those that are cut off, or must double back, either by tacking or wearing. Here Mr Clerk shows, that if the enemy follow the first of these methods, and make his line either tack in succession, or all together, such a distance must be left between them and the three or four sternmost ships, that not only must those last be easily carried, but that several more must probably be thrown into such a situation as to subject them almost unavoidably to the same fate. If the enemy attempt the same thing by wearing, his condition will be still worse. The fleet, by falling to leeward, must not only desert the ships attacked altogether, but must leave the sternmost of the wearing ships so much exposed as to render it certain that they will be entirely cut off.

At the time when this method of attack was proposed, it was regarded as a manœuvre quite new, and as having never yet been acted on. Mr Clerk, indeed, has entered into a historical detail, which tends to establish this point, and in which, from the most authentic documents, he traces the plans of most of our remarkable naval actions, from that of Admiral Matthews, off Toulon, in 1744, to that of Admiral Greaves off the Chesapeak in 1781. In most of these actions, though conducted by some of our ablest naval officers, the British fleet being to windward, by extending the line of battle, with a design of disabling or destroying the whole of the enemy's line to leeward, was itself disabled before the ships could reach a situation in which they could annoy the enemy; while, on the other hand, the French, perceiving the British ships in disorder, have made sail, and, after throwing in their whole fire, have formed a line to leeward, where they lay prepared for another attack; and thus has been frustrated that combination of skill and courage which distinguishes our seamen, and has always been so conspicuous in actions of single ships. The analysis of those actions forms a most interesting part of Mr Clerk's book, and furnishes a commentary on the naval history of Britain, such as we seek for in vain in the treatises written expressly on that subject.

In the second part of his work, which, though first written, was last published, the author has considered the nature of the attack from the leeward, or where the fleet which would force the other to action has not the advantage of the weather-gage. Here also he proves, by arguments very clear and convincing, that nothing promises success but the cutting of the enemy's line in two; the leeward fleet on the opposite tack to that of the enemy, bearing up obliquely, so as to pierce the linė in the centre, or towards the rear, as circumstances may direct. The ships thus cut off could have no support, and must either save themselves by downright flight, or fall into the hands of the enemy.

The time when Mr Clerk was engaged in these speculations, was a period very memorable in the naval, the military, and political history of this country; and never was there a moment when the communication of the secret he had discovered could have been attended with more important consequences. The contest in which Britain was engaged with the American colonies, so questionable in its principle,—so approved by the nation,—and so obstinately pursued by the Government, had involved the country in the greatest difficulties. A series of great and ill-directed efforts, if they had not exhausted, had so far impaired, the strength and resources of Britain, that neighbouring nations thought they had found a favourable opportunity for breaking the power, and humbling the pride, of a formidable rival. The French Government, desirous of accomplishing an object of which it had never lost sight, and willing also to share in the glory of giving independence to a new State, was yet ignorant of the lesson which it was so soon to learn to its cost, the danger which a despot runs, who attempts to give that liberty to other nations which he refuses to his own people. Spain also, which we see at this moment exerting every nerve to continue the thraldom of her own colonies, joined early in the scheme of giving independence to those of England; and by her detail of a hundred grievances, sufficiently convinced the world, that her hostility to Britain proceeded from a cause which she could not venture to avow.—Against this formidable combination, which Holland was preparing to join, Britain stood alone without an ally; and not merely alone, but divided in her counsels, with more than half her force engaged in the operations of a destructive civil war, in which victory would have been more ruinous than defeat. These were circumstances, which, in the mind of every friend to his country, could not but excite anxiety and alarm; yet they were perhaps not the most threatening that distinguished this perilous crisis. In the naval rencounters which took place after France had joined herself to America, the superiority of the British navy seemed almost to disappear; the naval armies of our enemies were every day gaining strength; the number and force of their ships were augmenting; the skill and experience of their seamen appeared to be coming nearer to an equality with our own. Their commanders were completely masters of the art of avoiding a general or decisive action, and at the same time of materially injuring their enemies. In the doubtful conflict off Ushant, which gave the commencement to our hostilities with France, the British admiral, after placing himself between the French fleet and their own coast, continued to manœuvre for several days together, without being able to bring on a general action, and was forced at length to draw off towards his own ports, allowing the French to return to theirs, without the capture of a single ship to support his own claim to victory, or to refute that of the enemy. The year which followed this had witnessed the most inglorious naval campaign recorded in the annals of Great Britain. The combined fleets of France and Spain were seen riding with exultation in the British Channel, capturing our ships close to our own shores, while the naval force of Britain stood aloof, and only ventured to look from a distance on a scene which every British seaman beheld with grief and indignation, while he seemed to read in it the tale of his personal dishonour. Another action in the course of the same year, had no great tendency to console us for the disgraceful caution which our fleet in the Channel had been forced to observe. Admiral Byron attempted to bring the French fleet, off Grenada, to action, and after the greatest gallantry, displayed both by himself and the officers under his command, he entirely failed in his object, and even suffered considerable loss. Indeed, when one studies the account of this action, by help of the light which the author of the *Naval Tactics* has thrown on it, he sees with much regret the highest efforts of valour and seamanship thrown away, from our ignorance of the true principle by which our attack should have been directed; while the French, in their position to leeward, succeeded, with their usual address, in damaging our ships, and saving their own.

The parallel drawn by Mr Clerk between the unfortunate engagement of Admiral Byng and this of Admiral Byron, is sufficiently striking, and shows but too clearly, that there are many circumstances, besides conduct and valour, that determine the character of a soldier that fights either at sea or land.

The action of Admiral Arbuthnot in the succeeding year, deceived equally the hopes of the nation, and equally demonstrated the skill of the French commanders, in the means of obtaining the end they had in view, and in entirely defeating that of their enemy; and by its unhappy influence on our military operations on shore, may be regarded as the most fatal miscarriage that marked the progress of the British arms. The action of Admiral Greaves off the Chesapeak, concluded a series of unsuccessful attempts, in which, though no signal disaster fell on the British fleet, no glory was gained, the ultimate object of the expedition was always lost, and, to a power used to boast in its superiority, the entire absence of victory seemed equivalent to defeat. The enemy was acquiring skill and confidence, while we were losing that feeling of superiority on which success so often depends. The circumstances of the nation had never called on every individual to think more seriously of the situation of his country; nothing had ever proved so clearly, that, at sea, the system of offensive warfare was yet but imperfectly understood, nor was there ever a juncture, when such discoveries as Mr Clerk had made, could be brought forward with so great effect. To a man who, like him, was a real lover of his country, sincerely interested in its liberty and independence, as well as in the glory of its flag; full also of the enthusiasm of genius, and the love of science; I can hardly imagine any higher or more exquisite delight, than that which he must have felt, when his imagination arose from the despondency into which the actual state of things had thrown every thinking man, to consider the change which the secret which he had in his possession was likely, nay sure to make, in the condition of his country.

There can exist, I think, but one feeling superior to this,—that which must arise on seeing this noble vision realized. This also fell to the share of the author of the *Naval Tactics*, who lived to see his measures carried into effect with unexampled skill and gallantry; saw them lead to victories more splendid than the most sanguine hopes could have ventured to anticipate, and saw himself become one of the great instruments by which Providence enabled his country to weather a more awful tempest than any by which it had hitherto been assailed.

Being fully satisfied as to the principles of his system, Mr Clerk had begun to make it known to his friends as early as 1779. After the trial of Admiral Keppell had brought the whole proceedings of the affair off Ushant before the public, Mr Clerk made some strictures on the action, which he put in writing, illustrating them by drawings and plans, containing sketches of what might have been attempted, if the attack had been regulated by other principles, and these he communicated to several naval officers, and to his friends both in Edinburgh and London.

In the following year he visited London himself, and had many conferences with men connected with the navy, among whom he has mentioned Mr Atkinson, the particular friend of Sir George Rodney, the Admiral who was now preparing to take the command of the fleet in the West Indies. A more direct channel of communication with Admiral Rodney was the late Sir Charles Douglas, who went out several months after the Admiral, in order to serve as his Captain, and did actually serve in that capacity in the memorable action of the 12th of April 1782. Sir Charles, before leaving Britain, had many conferences with Mr Clerk on the subject of naval tactics, and, before he sailed, was in complete possession of that system. Some of the conferences with Sir Charles were by the appointment of the late Dr Blair, prebendary of Westminster; and at one of these interviews were present Mr William and Mr James Adam, with their nephew the present Lord Chief Commissioner for Scotland. Sir Charles had commanded the Stirling Castle in Keppell's engagement; and Mr Clerk now communicated to him the whole of his strictures on that action, with the plans and demonstrations, on which the manner of the attack from the leeward was fully developed.

The matter which Sir Charles seemed most unwilling to admit, was the advantage of the attack from that quarter; and it was indeed the thing most inconsistent with the instructions given to all admirals.

Lord Rodney himself, however, was more easily convinced; and in the action off Martinico, in April 1780, the original plan seemed regulated by the principles of the *Naval Tactics*. The British fleet was to leeward, and the first signal made by the Admiral gave notice of his intention to attack the enemy's rear with his whole force. The enemy, however, having discovered this intention, wore, and formed on the opposite tack, and thus the effect of the signal was for the time defeated. The Admiral appeared then to depart from the new system; for the next signal which he threw out, was for every ship *to bear down* on her opposite, according to the 21st article of the additional fighting instructions. It appears, as we shall afterwards see, that the cause of this change was the mistake of the signals, the captains of the fleet not being sufficiently prepared for the new method of attack. In the two actions which immediately followed this, on the 15th and 18th of the next month, the French succeeded in the defensive system; and it was not till two years afterwards, in April 1782, that Lord Rodney gave the first example of completely breaking through the line of the enemy, and of the signal success which must ever accompany that manœuvre, when skilfully conducted. The circumstances were very remarkable, and highly to the credit of the gallantry as well as conduct of the Admiral. The British fleet was to leeward, and its van, on reaching the centre of the enemy, bore away as usual along his line; and had the same been done by all the ships that followed, the ordinary indecisive result would infallibly have ensued. But the Formidable, Lord Rodney's own ship, kept close to the wind, and on perceiving an opening near the centre of the enemy, broke through at the head of the rear division, so that for the first time the enemy's line was cut in two, and all the consequences produced which Mr Clerk had predicted.

This action, which introduced a new system, gave a turn to our affairs at sea, and delivered the country from that state of depression into which it had been thrown, not by the defeat of its fleets, but by their entire want of success.

It was in the beginning of this year that the *Naval Tactics* appeared in print, though for more than a year before, copies of the book had been in circulation among Mr Clerk's friends. Immediately on the publication, copies were presented to the Minister and the First Lord of the Admiralty. The Duke of Montague, a zealous friend of Mr Clerk's system, undertook the office of presenting a copy to the King.

Lord Rodney, who had done so much to prove the utility of this system, in conversation never concealed the obligation he felt to the author of it. Before going out to take the command of the fleet in the West Indies, he said one day to Mr Dundas, afterwards Lord Melville, "There is one Clerk, a countryman of yours, who has taught us how to fight, and appears to know more of the matter than any of us. If ever I meet the French fleet, I intend to try his way."

He held the same language after his return. Lord Melville used often to meet him in society, and particularly at the house of Mr Henry Drummond, where he talked very unreservedly of the *Naval Tactics*, and of the use he had made of the system in his action of the 12th of April. A letter from General Ross states very particularly a conversation of the same kind, at which he was present. " It is (says the General) with an equal degree of pleasure and truth, that I now commit to writing what you heard me say in company at your house, to-wit, that at the table of the late Sir John Dalling, where I was in the habit of dining often, and meeting Lord Rodney, I heard his Lordship distinctly state, that he owed his success in the West Indies to the manœuvre of breaking the line, which he learned from Mr Clerk's book. This honourable and liberal confession of the gallant Admiral, made so deep an impression on me, that I can never forget it; and I am pleased to think that my recollection of the circumstance may be of the smallest use to a man with whom I am not acquainted, but who, in my opinion, has deserved so well of his country."

As a farther evidence of the sentiments of the Admiral on a subject where they are of so much weight, I have to quote a very curious and valuable document, a copy of the First Part of the *Naval Tactics*, with Notes on the margin by Lord Rodney himself, and communicated by him to the late General Clerk, by whom it was deposited in the family library at Penicuick. The notes are full of remarks on the justness of Mr Clerk's views, and on the instances wherein his own conduct had been in strict conformity with those views. He replies in one place to a question which Mr Clerk had put, (published after the action in spring 1780,) of which mention has been already made, concerning the conduct of that action. The first signal of the Admiral, as we have already seen, was for attacking the rear with his whole force. The French, perceiving this design, wore, and formed on the opposite tack. This made it impossible immediately to obey the Admiral's signal, and the next that he made was for every ship to attack her opposite. Mr Clerk's question was, Why did Sir George change his resolution of attacking the rear, and order an attack on the whole line?—Sir George answers to this, That he did not change his intention, but that his fleet disobeyed his signals, and forced him to abandon his plan.

An anecdote which sets a seal on the great and decisive testimony of the Noble Admiral, is worthy of being remembered, and I am glad to be able to record it on the authority of a Noble Earl. The present Lord Haddington met Lord Rodney at Spa, in the decline of life, when both his bodily and his mental powers were sinking under the weight of years. The Great Commander, who had been the bulwark of his country, and the terror of her enemies, lay stretched on his couch, while the memory of his own exploits seemed the only thing that interested his feelings, or afforded a subject for conversation. In this situation, he would often break out in praise of the *Naval Tactics*, exclaiming, with great earnestness, " John Clerk of Eldin for ever."

Generosity and candour seemed to have been such constituent elements in the mind of this gallant Admiral, that they were among the parts which longest resisted the influence of decay.

Soon after the victory obtained by Lord Rodney, the American war was brought to a conclusion, and the world enjoyed some years of repose. The French Revolution disturbed the tranquillity of Europe; Britain was quickly involved in a war more formidable than that in which the principles of Mr Clerk's system were first essayed; one where it was yet to be more severely tried, and was yet to render more important services to the country.

We have seen, that Lord Rodney had been so convinced by the first explanation he received of Mr Clerk's system, that he declared, that should he meet the French fleet, " he would try his way." —On Lord Howe, the effect of the first perusal of the same work was quite different, though the result in the end was entirely the same. A copy of the first edition of the *Naval Tactics* was sent to his Lordship, who, after reading it, expressed himself as highly pleased with the ingenuity of the book, and as greatly struck with the circumstance of the author being a landsman; but he nevertheless desired General Clerk to inform his ingenious kinsman, that he would adhere to the old system if ever he had an opportunity of engaging the French fleet. To this Mr Clerk replied, through the same channel, that if his Lordship did so, he would infallibly be beaten. His Lordship, however, when it came to the trial, did not adhere to the old system, but, concentrating his force, directed it against one point, precisely on the principles of the *Naval Tactics*. His change of opinion may have arisen from the practical commentary by which Lord Rodney had illustrated the principles of that work; and perhaps, too, a second perusal of the book itself had materially contributed to this effect. That Lord Howe really consulted it a second time, there is good reason to believe. When he commanded the Channel fleet in 1793, Mr James Clerk, the youngest son of the author of the *Naval Tactics*, a young man of great promise, who, had he lived, would have done honour to the profession on which his father had bestowed so valuable a gift, served as a midshipman on board the Admiral's ship the Queen Charlotte. He possessed a copy of the second edition of his father's book, which was borrowed by Captain Christian, no doubt for the Admiral's use. Thus much is certain, that the action of the 1st of June 1794, was, in its management, quite conformable to Mr Clerk's system, and its success entirely owing to the manœuvre of breaking the line.

Lord Howe was also the first who introduced into the signal book signals directed to the object of cutting off the rear,—of bringing the whole force to bear on one point,—breaking the line, &c. Indeed, if his Lordship's conduct had been contrary to the principles of the *Naval Tactics*, the words of his declaration, that he would still adhere to the old method, is a decided testimony in favour of one of the points which I think it most material to establish. About the utility of the method, after Lord Rodney's action, no doubt could be entertained. As to its novelty, and its originality, if any difference of opinion could arise, it is completely answered by Lord Howe's message delivered to General Clerk, as it is a proof that an officer of his Lordship's great skill and experience, considered this manœuvre as new, as opposed to the ordinary practice, and as a thing hitherto unknown. The novelty of the system, therefore, can no more be doubted than its utility.

An example of breaking the line, with success, if possible, more brilliant than either of the preceding instances, was afforded by Lord St Vincent's memorable action on the coast of Spain, when, disregarding, as he said, in his own account of it, the regular system, he attacked the Spanish line of twenty-seven ships, with fifteen only, and by carrying a press of sail, intersected and cut off the windward division, of which four were taken before the rest of the fleet could work up to their relief.

Lord St Vincent had early been made acquainted with Mr Clerk's book, of which a copy had been sent him by Colonel Debbieg of the Royal Engineers, a particular friend of the author. I do not find that his Lordship ever expressed any opinion on the principles of this work.

Lord Duncan's victory on the coast of Holland was achieved on the same principle, and carried into effect with singular gallantry. His Lordship, indeed, before going to sea, had many conferences with Mr Clerk, and professed that he was determined to pursue the plan of operations which he had pointed out. His Lordship's attack, accordingly, was directed against the centre of the enemy, in consequence of which the rear division was cut off and taken, with the exception of a single ship. When the first news of this victory, so near to our own shores, and therefore so strongly felt, and so highly appreciated by us all, was brought to Walmer Castle, where Mr Pitt was then residing, he, with Lord Melville, Mr Fordyce, and some others, were sitting at table just after dinner. A man who had seen the action, and had just landed, desired to be introduced, and on coming into the room, gave an account of what he had witnessed; on his mentioning that Lord Duncan had broke through the Dutch line, Lord Melville immediately exclaimed, Here is a new instance of the success of Clerk's system.

The last and greatest in the brilliant series of victories that followed the publication of the *Naval Tactics*, was, like all the rest, obtained by the skilful application of the principles there unfolded; and of this, Lord Nelson's instructions, before the battle, are the fullest evidence. They even contain, in the body of them, several sentences that are entirely taken from the *Naval Tactics*. These instructions were transmitted to Mr Clerk by one of the commanders in that memorable action, Captain, now Admiral Sir Philip Durham, with a note, which shows in what light his discoveries were viewed by those most capable to decide. " Captain Durham, sensible of the many advantages which have accrued to the British nation from the publication of Mr Clerk's *Naval Tactics*, and particularly from that part of them which recommends breaking through the enemy's line, begs to offer him the inclosed form of battle, which was most punctually attended to in the brilliant and glorious action of the 21st of October. Mr Clerk will perceive with pleasure, that it is completely according to his own notions, and it is now sent as a token of respect from Captain Durham, to one who has merited so highly of his country.

"*H. M. S. off Cadiz, 29th Oct.* 1805."

I must observe, that the great Admiral, to whose last and most glorious action I have now alluded, had put in practice the same manœuvre in the battle of the Nile; the line was then broken in the same way, and the discomfiture, by that means, of a fleet at anchor, was the most complete that can be imagined.

From the whole of this narrative, therefore, it is plain, that the *Naval Tactics* was acknowledged by professional men, as an original and valuable work, unfolding a new system; the advantages of which were proved by demonstrations founded on the most undeniable principles, and now verified by a series of great and brilliant victories, in consequence of which there has been effected an entire revolution in the offensive part of naval warfare. These truths having been so generally acknowledged and admitted, both by Naval Officers of the highest reputation, and by Statesmen of the greatest power, it cannot but appear extraordinary, that no mark of public favour was ever bestowed on the author, nor any acknowledgment made by Government of merit so distinguished. It was merit of the kind most directly calculated to interest the feelings, and to call forth the gratitude of the nation; it was an improvement in the art which Britain reckons so peculiarly her own; it was a contrivance for making more effectual the arms in which she most confides; for rendering more impenetrable the wooden walls, to which she trusts her safety, her prosperity, and her independence. The name of Mr Clerk, and of the *Naval Tactics*, is in the mouth of every officer, from the midshipman to the admiral.

Whatever was the cause of this strange omission, it is deeply to be regretted,—regretted, however, much less on account of Mr Clerk, than on account of the nation itself. To a man conscious of having rendered so important a service to his country as he had done,—who might say to himself without vanity, that he was entitled to be numbered with her most useful citizens, and her most eminent benefactors,—who saw that the actions which had immortalized the names of Rodney, Howe, Duncan, and Nelson, had been all directed by a principle which he had been the first to discover,—and who knew, that he was to go down to posterity as the author of a great and important improvement;—to the happiness of a mind sustained by such reflections, and inspired by the sentiments which must accompany them, what great addition is it in the power of a monarch, or even of the nation, to make? what is it that the common badges and titles of honour and distinction can be supposed to add? These may be fit, and even necessary emblems, for marking degrees of merit of an ordinary kind; but when merit is transcendant to a certain point, it can dispense with such conventional symbols; it shines of its own light, and enables its possessor to look down on the neglect or the ingratitude of the world.

But though these considerations may in some measure set us at ease with respect to the author himself, and his own feelings, it must be allowed that they take nothing from the blame incurred by those to whom the nation had intrusted the power of dispensing its honours and rewards. Neglect of merit will always operate as a discouragement to exertion, and every instance of it tends to extinguish a portion of the fire of genius, of that which often constitutes the sole riches of the possessor, and is always a valuable portion of the patrimony of the state. Every mind is not provided with the power of enduring neglect; ingenious men are often the most sensible of it; and it is hard that the possession of talents should be converted into a source of suffering. If the author of the *Naval Tactics* had not been supported by such enlarged views, and such high sentiments as we have mentioned, the circumstances of his case would have pressed on him with much severity.

That it was not ignorance of the facts, or of the chain of evidence now brought forward, that prevented a public acknowledgment of Mr Clerk's services, is altogether certain. The late Lord Melville, who held so conspicuous a situation in the government of this country during the greater part of the period I have been treating of, was early made acquainted by Mr Clerk himself with his ideas on the subject so often mentioned; and in the beginning of his political career, when yet King's Advocate in Scotland, was consulted on the best means of bringing forward those ideas, and gave his advice with the readiness and frankness for which he was remarkable in all situations of life. It is apparent, that he never ceased to hold Mr Clerk's discovery in the highest estimation; and of this, his observation at Walmer Castle (on hearing of Lord Duncan's victory above related) is a sufficient proof,—an observation that conveyed a severe censure on himself, and on the minister to whom it was addressed, unless they both felt that their power of rewarding the merit in question was restrained by some considerations known to themselves, and invisible to the public at large.

Lord Melville had particularly studied the affairs and the interests of the Navy; he had been for a long time at the head of the Admiralty; and there is reason to think, that he was sensible of the improper neglect with which the author of the *Naval Tactics* had been treated. I have been assured that he had represented this to Mr Pitt, but when it was too late, and when that minister was drawing near the end of life.

If I might venture any conjecture on the cause of an omission which it is impossible to justify, I should be disposed to ascribe it to the fear of giving offence to the Navy, and to consider it rather as resulting from an excess of caution, than from direct or intentional neglect. It might seem to derogate from the glory of our Naval Officers, to recognise a landsman as the author of one of the most valuable discoveries that had been made in their own art,—as the person who had not only pointed out the new principle, but had completely unfolded its advantages, and predicted its effects. If this were the ground on which the reward was withheld, it must at once be considered as very insufficient for the purpose of justification. The man entrusted with the power of rewarding merit, ought no more, than those who have committed to them the office of punishing guilt, to be accessible to any voice but that of truth and justice. The little and mean jealousies that may be excited, by an impartial discharge of their duty, ought never to interfere with the performance of what is imperiously called for. Jealousy, in the present instance, was a weakness that deserves no indulgence; it was vanity and selfishness that ought to have met with no sympathy, no toleration. If, indeed, such feelings any where exist, there is fortunately no reason to think them general; and it is a duty which I most willingly discharge, to say, that the Naval Officers with whom I have had the honour to converse on this subject, have all in the most unequivocal terms expressed their conviction of the importance and originality of Mr Clerk's discovery. That there are exceptions to this rule, I can only state as a conjecture, necessary to explain what is otherwise so difficult to be accounted for.

But to whatever cause the neglect of which I now complain is to be attributed, it is certain that the government and the navy have both lost a great opportunity of doing honour to themselves. A National Monument, that would have marked the era of this great improvement, and testified the gratitude of the nation to the author, would have been very creditable to the Minister under whose patronage it was erected; and an acknowledgment from the navy, that this discovery was the work of a landsman, would have been highly becoming in a profession, of which intrepidity and valour are not more characteristic than frankness and generosity.

END OF VOLUME THIRD.

Printed by George Ramsay & Co.
Edinburgh, 1822.

PLATE I.

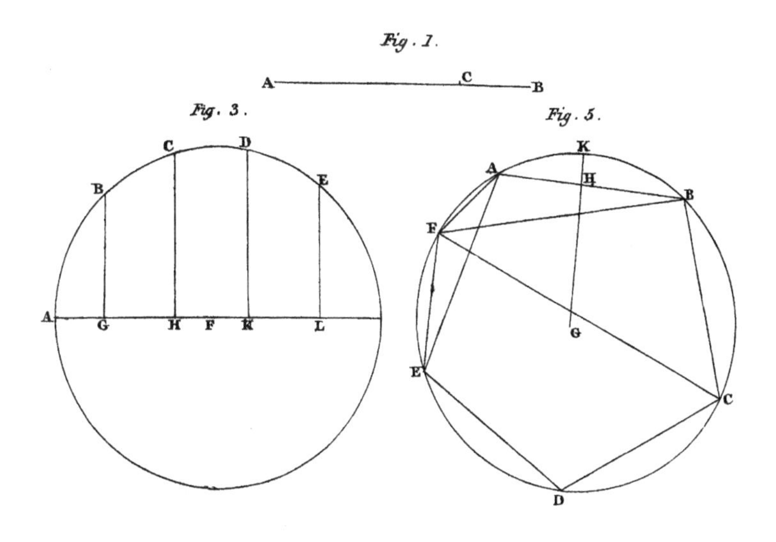

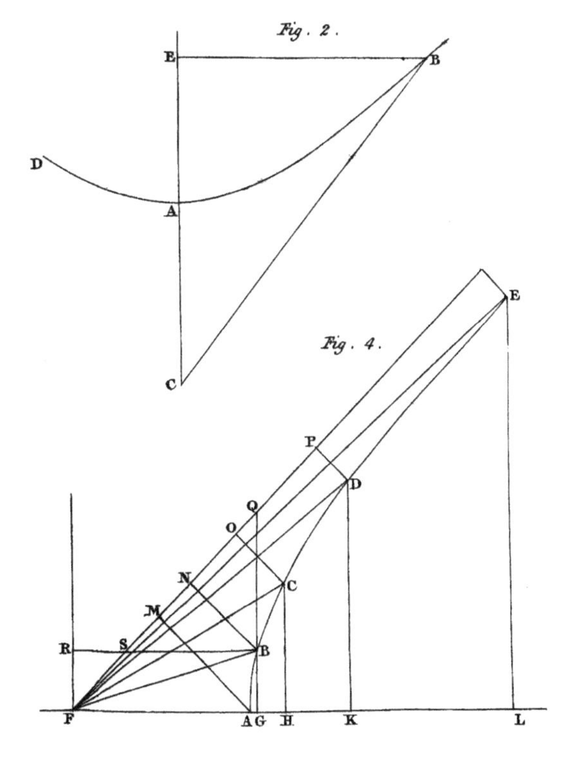

W. H. Lizars Sculpt.

PLATE II.

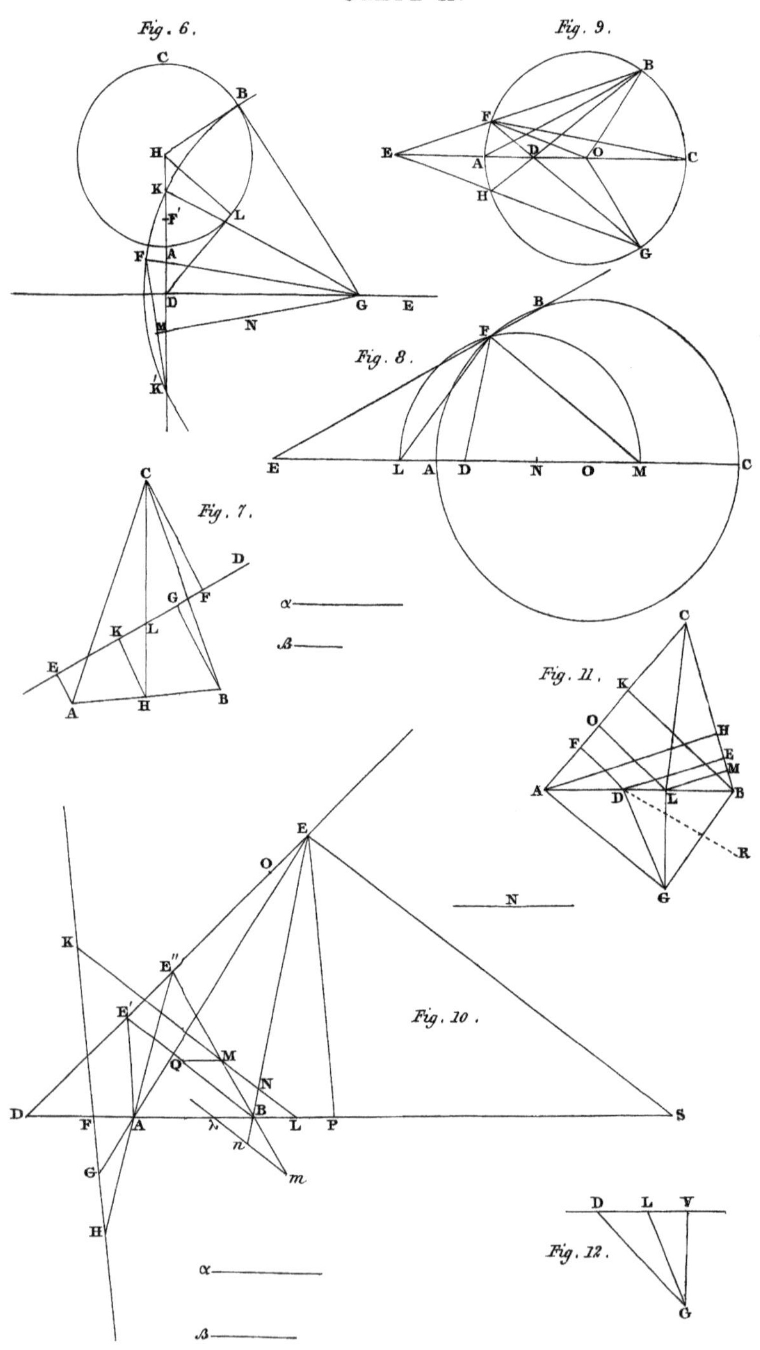

W.H. Lizars Sculp^t

PLATE III.

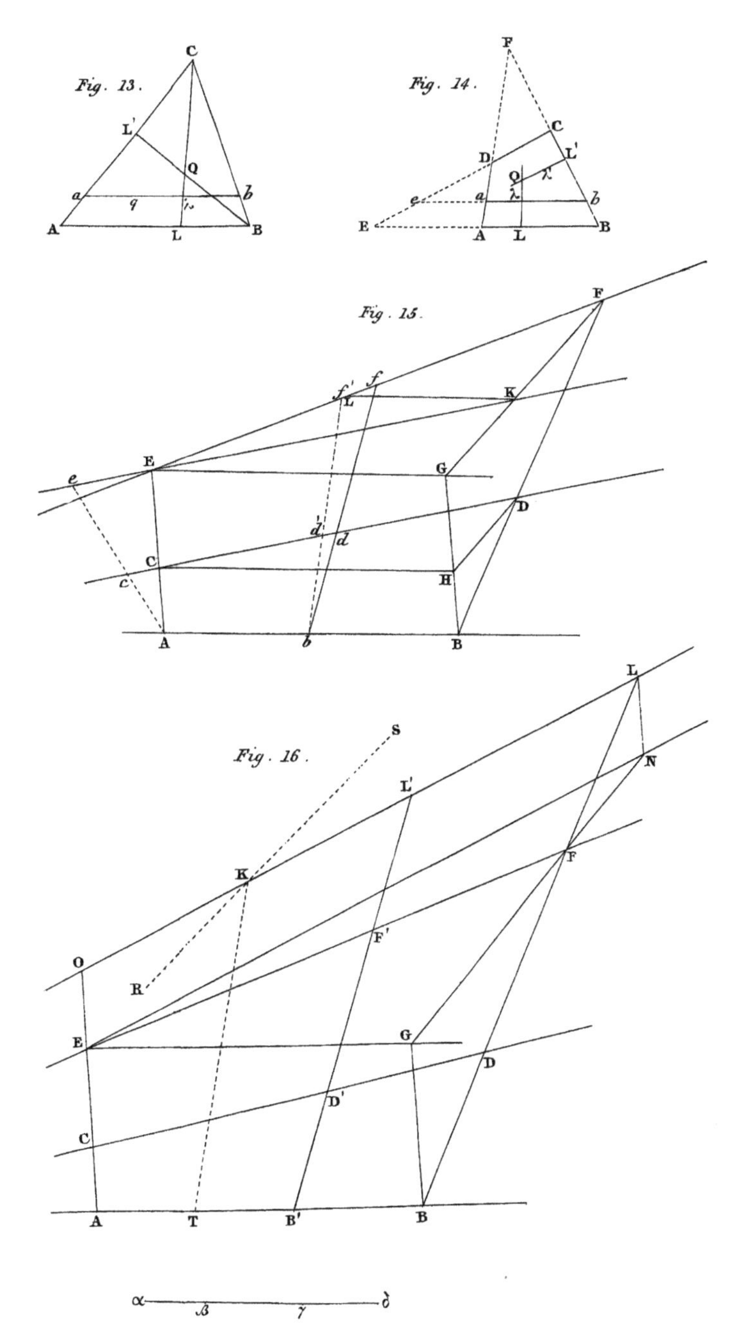

W. H. Lizars Sculp.t

PLATE IV.

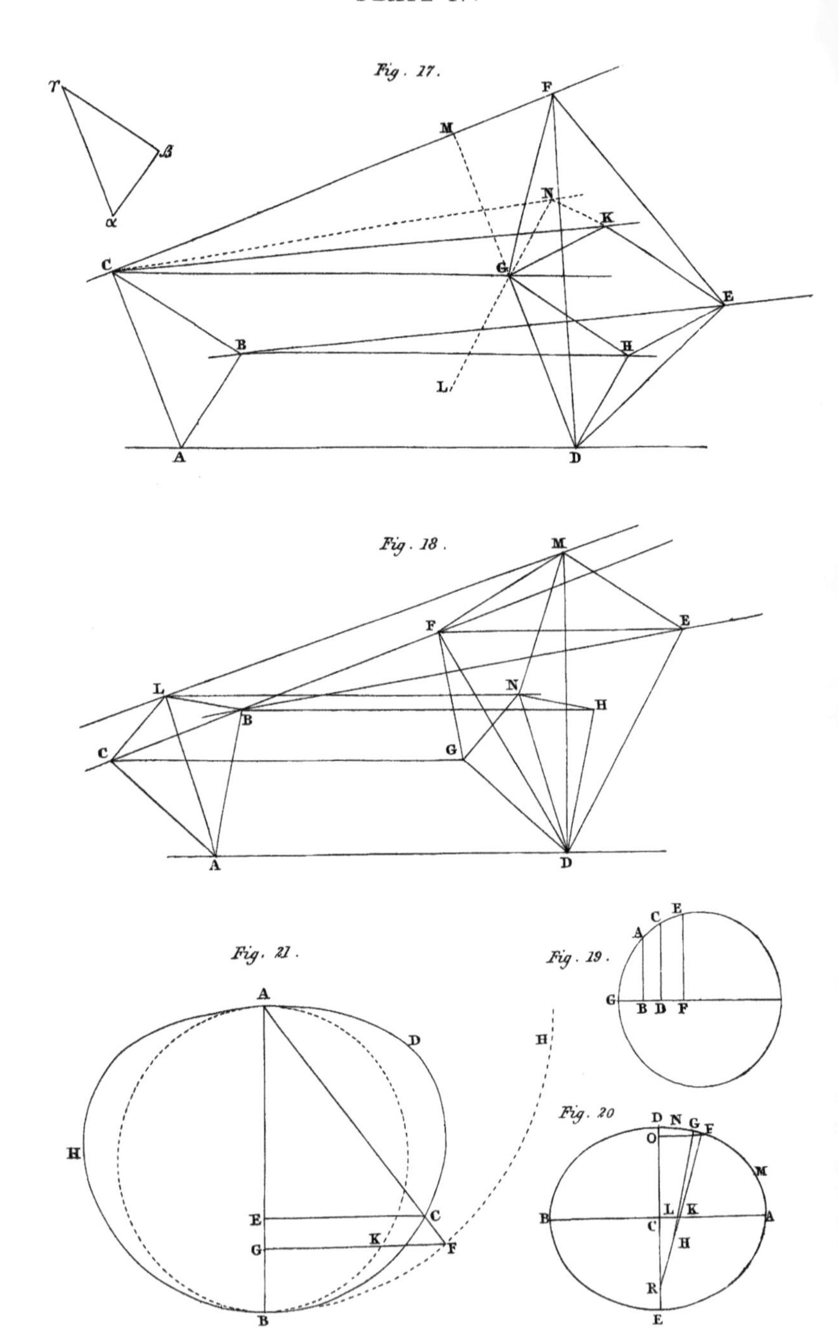
Fig. 17.
γ
β
α
F
M
N
K
C
G
E
B
H
L
A
D
Fig. 18.
M
F
E
L
N
H
B
C
G
A
D
Fig. 21.
A
D
H
H
E
C
G
K
F
B
Fig. 19.
E
C
A
G
B
D
F
Fig. 20
D
N
G
F
O
M
B
L
K
A
C
H
R
E
W.H. Lizars Sculpt

PLATE V.

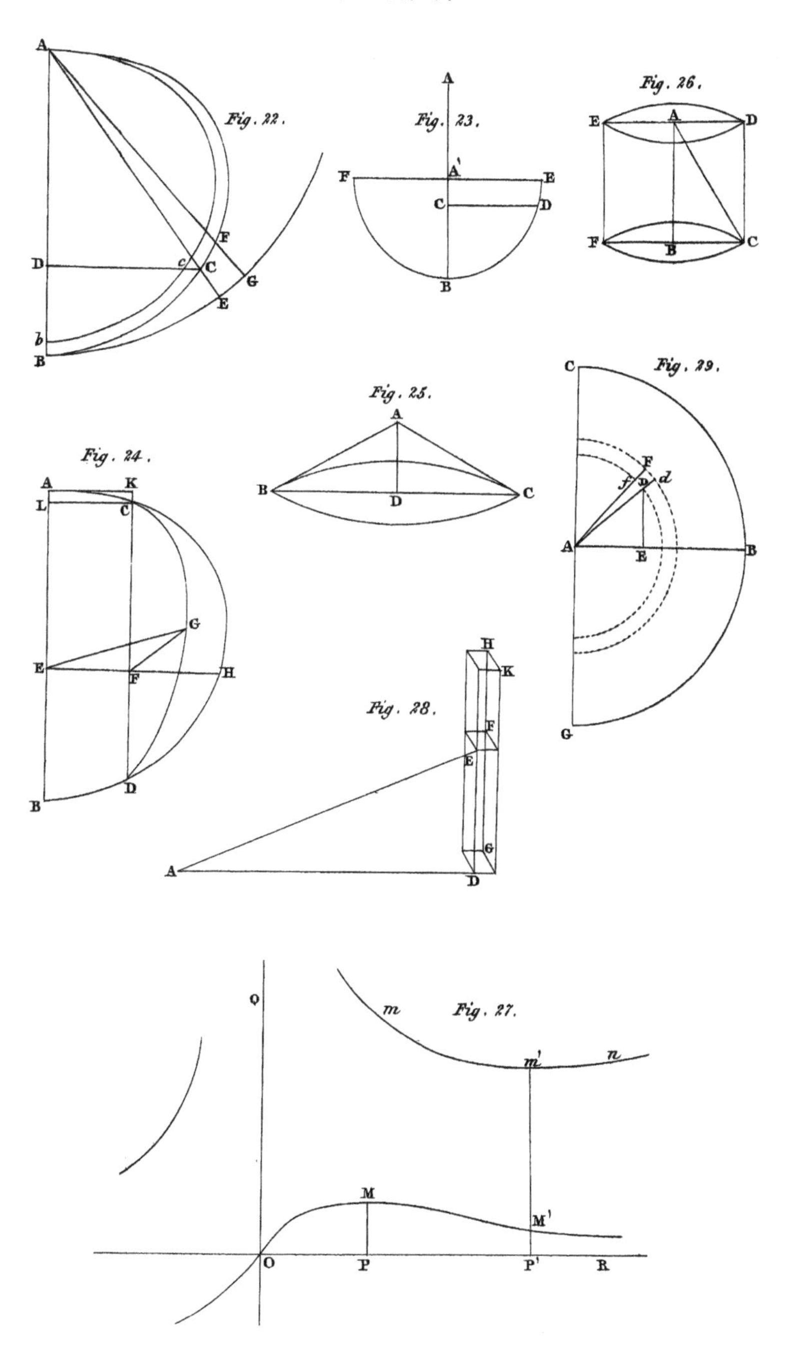

W.H. Lizars Sculpt

PLATE VI.

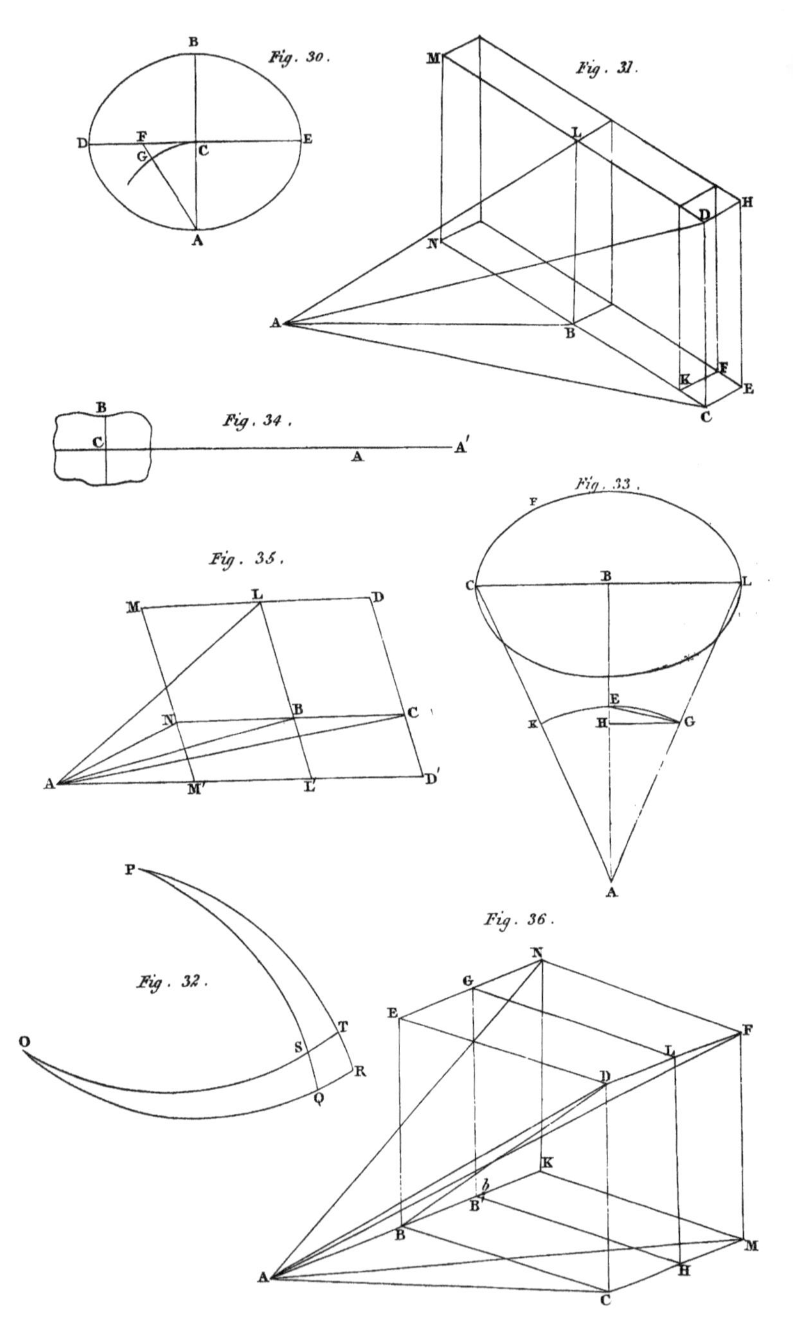

W. H. Lizars Sculp.t

PLATE VII.

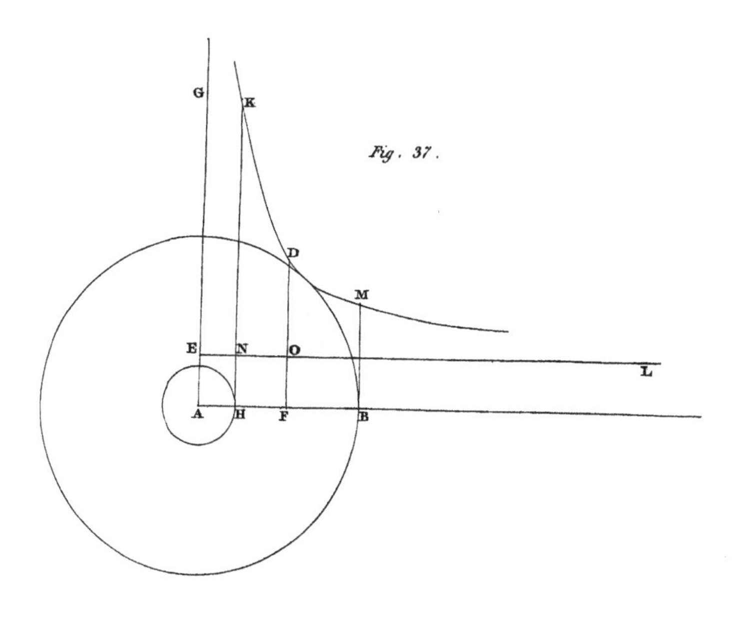

Fig. 37.

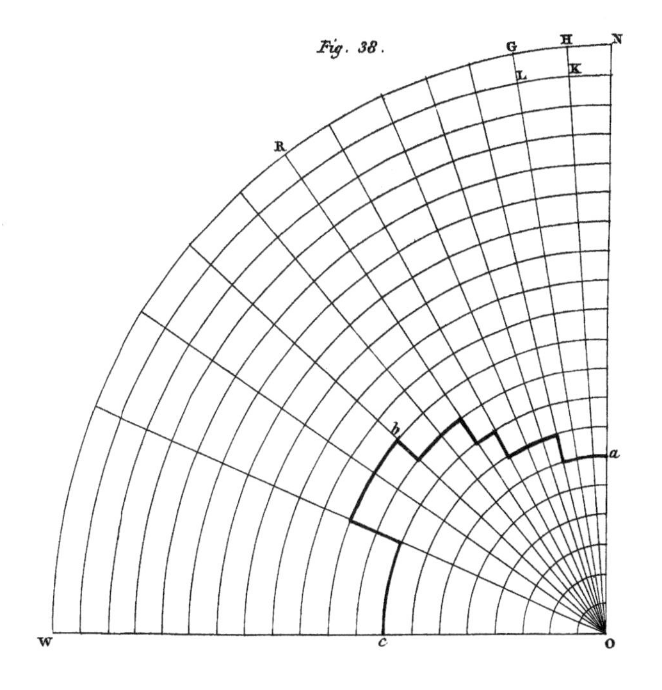

Fig. 38.

W.H. Lizars Sculp.t

PLATE VIII.

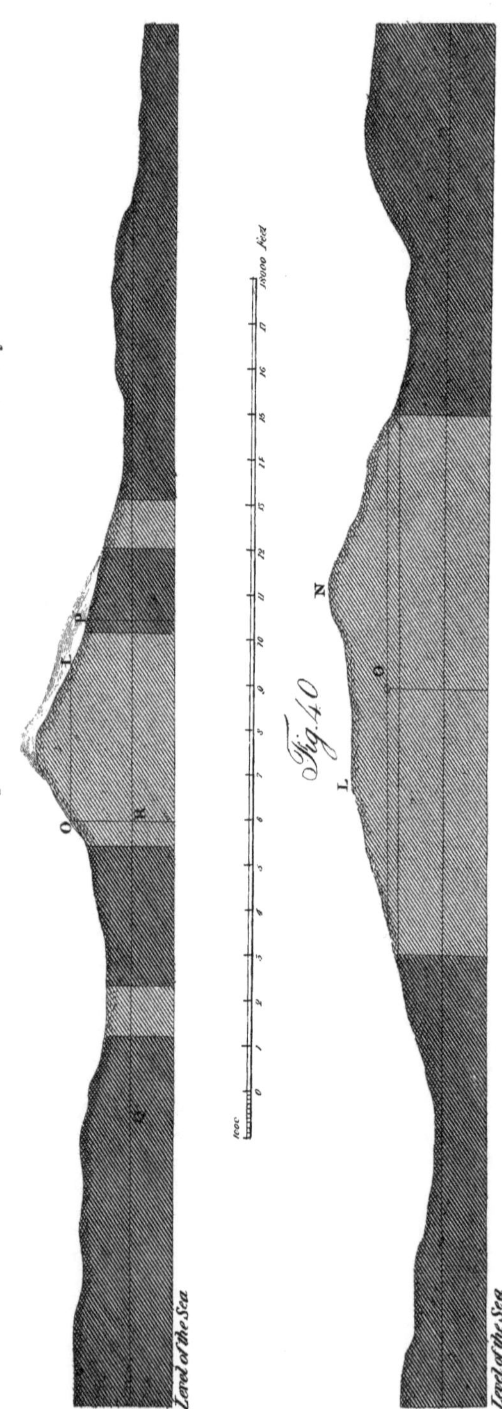
Fig. 39
Section of Schehallien in the plane of the Meridian of O the South Observatory
O
L
P
R
Q
Level of the Sea
1000
0
1
2
3
4
5
6
7
8
9
10
11
12
13
14
15
16
17
18000 feet
Fig. 40
N
O
L
Level of the Sea
Longitudinal section of Schehallien through the two Cairns N & K bearing from the Meridian N85.18½W.
W. H. Lizars Sculpt.